SpringerBriefs in Applied Sciences and Technology

Manufacturing and Surface Engineering

Series Editor

Joao Paulo Davim , Department of Mechanical Engineering, University of Aveiro, Aveiro, Portugal

This series fosters information exchange and discussion on all aspects of manufacturing and surface engineering for modern industry. This series focuses on manufacturing with emphasis in machining and forming technologies, including traditional machining (turning, milling, drilling, etc.), non-traditional machining (EDM, USM, LAM, etc.), abrasive machining, hard part machining, high speed machining, high efficiency machining, micromachining, internet-based machining, metal casting, joining, powder metallurgy, extrusion, forging, rolling, drawing, sheet metal forming, microforming, hydroforming, thermoforming, incremental forming, plastics/composites processing, ceramic processing, hybrid processes (thermal, plasma, chemical and electrical energy assisted methods), etc. The manufacturability of all materials will be considered, including metals, polymers, ceramics, composites, biomaterials, nanomaterials, etc. The series covers the full range of surface engineering aspects such as surface metrology, surface integrity, contact mechanics, friction and wear, lubrication and lubricants, coatings an surface treatments, multiscale tribology including biomedical systems and manufacturing processes. Moreover, the series covers the computational methods and optimization techniques applied in manufacturing and surface engineering. Contributions to this book series are welcome on all subjects of manufacturing and surface engineering. Especially welcome are books that pioneer new research directions, raise new questions and new possibilities, or examine old problems from a new angle. To submit a proposal or request further information, please contact Dr. Mayra Castro, Publishing Editor Applied Sciences, via mayra.castro@springer.com or Professor J. Paulo Davim, Book Series Editor, via pdavim@ua.pt.

Panagiotis Kyratsis · Anastasios Tzotzis ·
João Paulo Davim

CAD-based Programming
for Design and Manufacturing

 Springer

Panagiotis Kyratsis ⓘ
Department of Product and Systems Design
Engineering
University of Western Macedonia
Kozani, Greece

Anastasios Tzotzis
Department of Product and Systems Design
Engineering
University of Western Macedonia
Kozani, Greece

João Paulo Davim ⓘ
Department of Mechanical Engineering
University of Aveiro
Aveiro, Portugal

ISSN 2191-530X   ISSN 2191-5318 (electronic)
SpringerBriefs in Applied Sciences and Technology
ISSN 2365-8223   ISSN 2365-8231 (electronic)
Manufacturing and Surface Engineering
ISBN 978-3-031-78746-1   ISBN 978-3-031-78747-8 (eBook)
https://doi.org/10.1007/978-3-031-78747-8

This Springer imprint is published by the registered company Springer Nature Switzerland AG
The registered company address is: Gewerbestrasse 11, 6330 Cham, Switzerland

If disposing of this product, please recycle the paper.

Preface

The integration of Computer-Aided Design (CAD) and manufacturing technologies has reshaped the landscape of modern production, driving unprecedented levels of precision, efficiency, and innovation across industries. From automotive and aerospace, to consumer goods and medical devices, CAD-based programming has become a cornerstone of contemporary manufacturing, transforming how design and production of complex components and assemblies are conceptualized. In an era where technological advancements such as automation, digital twins, and additive manufacturing are pushing the boundaries of the design and manufacturing, mastering CAD-based programming is a necessity.

This book aims to provide a comprehensive exploration of CAD-based programming, specifically for design and manufacturing, serving as both an educational resource and a practical guide for designers and engineers. Irrespectively of the reader's background and expertise, this book can offer valuable insights into the tools, methodologies, and best practices that define CAD-based workflows in today's industry.

Throughout the five chapters, topics such as the automated design of products, the geometry feature recognition, the automated assembly of systems, and the CNC code generation from CAD model are covered. In addition, the integration of CAD with CAM and other modules is discussed, as well as the applications of CAD-based programming, the challenges and the opportunities presented by emerging technologies like artificial intelligence and machine learning, as they continue to shape the future of CAD-based programming.

Finally, this book aims to contribute toward the deeper understanding of CAD-based programming, but also inspire the reader to explore new possibilities within the dynamic intersection of design and manufacturing.

Kozani, Greece   Panagiotis Kyratsis
Kozani, Greece   Anastasios Tzotzis
Aveiro, Portugal   João Paulo Davim

Contents

Chapter 1
A Review on CAD-Based Programming for Design and Manufacturing

Abstract Integration of Computer-Aided Design (CAD) with Computer-Aided Manufacturing (CAM) is one of the most researched topics regarding modern manufacturing. Programming of CAD systems by utilizing the Application Programming Interface (API) and similar resources has become increasingly necessary in the past years, for the automation of routine engineering tasks, the development of tools and the analysis of complex processes. In the present paper, an effort was made to pinpoint the advantages and disadvantages that derive from the use of CAD-based programming, especially in the areas of design and manufacturing. For this purpose, a variety of relevant research works from the last decade has been processed and analyzed, with an aim to highlight the usefulness of CAD-based programming and its applications. In addition, the challenges and future trends have been identified as well. To visualize the research, the gathered research papers have been classified in terms of the publication year, research area, and application groups.

Keywords Application programming interface (API) · Automated assembly · Automated design · Computer-aided design (CAD) · Computer-aided manufacturing (CAM) · CAD-based design · CAD-based manufacturing · SolidWorks™

1.1 Introduction

Computer-Aided Design (CAD) has revolutionized the fields of engineering, architecture, and manufacturing by enabling precise, efficient, and flexible design processes. While traditional CAD software allows users to create detailed 2D and 3D models, CAD-based programming takes this a step further by integrating scripting languages and algorithms into the design process. This integration allows for the automation of routine or advanced tasks, the creation of parametric models, and the development of custom design tools, significantly enhancing the capabilities of traditional CAD systems.

CAD-based programming involves using programming languages to automate and extend the functionality of CAD software. This can be achieved through scripting languages like Python, Visual Basic for Applications™ (VBA), or proprietary languages specific to CAD platforms, such as AutoLISP for AutoCAD. These scripts enable designers to automate repetitive tasks, generate complex geometries, and customize their CAD environment to meet specific project requirements. One of the key aspects of CAD-based programming is parametric design, where models are defined by parameters such as dimensions, angles, and constraints. By adjusting these parameters, designers can quickly generate different design variations without manually redrawing each element. This flexibility is particularly valuable in fields where design customization is critical, such as in product design or architectural modeling.

The primary benefit of CAD-based programming is increased efficiency. Automating repetitive tasks and streamlining the design process saves time and reduces the potential for human error. For example, a script can be used to automatically generate a series of holes along a surface or create a complex pattern that would be time-consuming to model manually. This not only accelerates the design process but also ensures consistency and precision across all design iterations. Another significant advantage is the ability to explore a broader range of design possibilities through algorithmic design. By incorporating algorithms into the design process, engineers, and architects can explore numerous design options quickly, optimizing for factors such as material usage, structural integrity, or aesthetic appeal. This is particularly evident in generative design, where software generates a vast number of potential designs based on specific criteria, allowing designers to select the most effective solution. Additionally, CAD-based programming allows for greater customization of the CAD environment itself. Designers can create custom tools and interfaces that align with their specific workflow needs, enhancing productivity and ensuring that the software works seamlessly with other specialized tools, such as Computer-Aided Manufacturing (CAM) or Finite Element Analysis (FEA).

While CAD-based programming offers numerous advantages, it also presents challenges. The most notable is the learning curve associated with mastering programming languages and understanding how to integrate them effectively into the CAD environment. Designers who are accustomed to traditional CAD methods may find the transition to CAD-based programming daunting. Moreover, developing scripts and algorithms requires a deep understanding of both design principles and programming logic, which can be time-consuming to acquire. However, once these skills are developed, the benefits in terms of efficiency, flexibility, and innovation are significant.

As technology continues to advance, the integration of CAD with programming will play an increasingly critical role in shaping the future of design. In the present review, an effort was made to collect the points that require attention in the areas of automated design and manufacturing via CAD-based programming. Specifically, the review focuses on the recent state of the art, the applications, the benefits and the challenges that emerge from the use of CAD-based programming in the aforementioned areas.

1.2   State of the Art

1.2.1   Automated Design Based on CAD Programming

Automated design based on CAD programming represents a paradigm shift in the fields of engineering and manufacturing. By integrating programming with CAD, designers can automate complex tasks, create parametric models, and develop custom tools that enhance productivity and precision. This approach leverages the power of algorithms to extend the capabilities of traditional CAD systems, enabling more efficient and innovative design processes.

The concept of automating design tasks within CAD systems has evolved along-side advancements in computational power and software development. Early CAD systems, which emerged in the 1960s, were primarily used for drafting and simple 2D representations [1]. However, as CAD software became more sophisticated, the need for automation grew, leading to the integration of scripting languages and APIs that allow users to customize and automate design processes. One of the early milestones in this evolution was the introduction of parametric modeling in the 1980s, which allowed designers to define relationships between different elements of a design [2]. This capability laid the foundation for more advanced automation techniques, such as generative design and algorithmic modeling, which have become increasingly prevalent in recent years. Automated design based on CAD programming involves the use of scripting languages, APIs, and algorithms to create, manipulate, and opti-mize CAD models. The core principles of this approach include parametric design, scripting and customization, algorithmic design and integration of CAD with other systems.

Parametric design is a key principle of automated CAD programming. It involves creating models that are defined by parameters, such as dimensions, angles, and constraints, which can be adjusted to generate different design variations. By defining these relationships programmatically, designers can easily modify their models and explore a wide range of design possibilities without manually redrawing or adjusting each component [2]. As already mentioned in the introduction, CAD programming allows users to write scripts in languages such as Python, VBA™, or proprietary languages specific to CAD software (e.g., AutoLISP for AutoCAD). These scripts can automate repetitive tasks, generate complex geometries, and customize the CAD environment to suit specific design needs. Scripting is particularly useful for tasks that would be time-consuming or error-prone if done manually [3]. Moreno and Bazán [4] studied the advantages that derive by using traditional techniques of technical drawing in processes for automation of the design, with CAD systems that are nonparametric, provided with scripting languages. Another study by Marschallinger et al. [5] verified the importance of scripting in CAD-related tasks. The authors have developed a voxelizer for solid models with Visual LISP.

Algorithmic design involves the use of algorithms to generate or optimize CAD models. This approach can range from simple geometric transformations to complex generative design techniques, where algorithms explore a vast design space to identify optimal solutions based on predefined criteria. Generative design, in particular, has gained popularity for its ability to produce innovative and efficient designs that might not be achievable through traditional methods [6]. Automated CAD design often involves integrating CAD software with other tools and technologies, such as CAM, FEA, 3D printing and other types of software. The integration of CAD systems with spreadsheet software is shown in the study by Gella-Marín et al. [7]. A study related to FEA and CAD-based programming is evident in the study by Tzotzis et al. [8]. This integration allows for seamless workflows, where designs can be automatically optimized for manufacturability, tested for performance, and prepared for production with minimal manual intervention [9].

Applications of Automated Design Based on CAD Programming. In engineering, automated design is used to generate complex, parametric models of buildings, structures and products. Scripting languages, such as RhinoScript or Grasshopper for Rhino, allow engineers to create intricate facades, optimize structural components, and automate the creation of construction documents. For example, parametric modeling can be used to design a building's facade that adapts to different environmental conditions, optimizing sunlight exposure and energy efficiency [10]. A number of works [11–13] that engage the aforementioned scripting languages during the design process, reveal the significance of CAD-based programming, in the product design process as well.

Mechanical engineers use CAD programming to design and optimize components, assemblies, and systems. Scripting can automate the creation of complex geometries, such as gears or turbine blades, which would be difficult and time-consuming to model manually. Additionally, generative design algorithms can explore thousands of design iterations to find the most efficient solution for weight, strength, and manufacturability, which is particularly useful in industries such as aerospace and automotive engineering [14]. Junk and Burkart [15] compared different CAD systems for generative design purposes within the 3D printing industry. Bacciaglia et al. [16] employed the Boolean operations of a CAD system to subdivide large-scale components for facilitating the manufacturing process. This method can be applied for aerospace and automotive parts, fabricated with the additive technology. Forrai et al. [17] developed a tool in CATIA™ that focuses on the geometrical definition and the geometrical features reusability, with an aim to create robust parametric shape-based CAD models. Similar techniques and methods have been applied in other works [17–20] related to the aerospace and automotive industry as well.

In product design, CAD programming is used to create parametric models that can be easily customized for different markets or consumer preferences. For example, a company producing custom-fit products, such as eyewear or footwear, can use parametric design to adjust the shape and size of their products based on individual customer measurements. This approach not only speeds up the design process but also ensures a perfect fit and better customer satisfaction [9]. Dalpadulo et al. [21] worked on the testing of the effectiveness of CAD-based tools with respect to design and optimization of additively manufactured parts. The paper focuses on the automotive industry. Kyratsis et al. [22] employed the API of SolidWorks™ to develop a convenient tool with graphical interface, for the automated design of a standardized product. The tool combines the capabilities of a commercially available CAD system with the programming resources embedded to the system. In the same manner, the next studies [23–25] implement CAD-based tools for the design, verification, and optimization of product systems.

In industrial design, CAD programming facilitates the creation of products that are both functional and aesthetically pleasing. Designers can use algorithms to explore a wide range of shapes and forms, optimizing them for both performance and manufacturability. In manufacturing, CAD programming is used to automate the generation of toolpaths for CNC machining, ensuring that parts are produced with precision and efficiency. Kukreja et al. [26] developed an algorithm for the CNC toolpath planning of complex parts, by using voxel-based CAD models, focusing on the zig-zag toolpath planning. Favi and Campi [27] worked on a method for integrating design for welding method with CAD systems. This CAD-based tool can be used to avoid design problems related to the welding of components. The integration of CAD with CAM software allows for a seamless transition from design to production, reducing lead times and improving product quality [9]. Integration of CAD with CAM is further discussed in Sect. 2.2. Matta et al. [28] studied the possibility of integrating CAD with CAM systems for remote parts manufacturing for rapid prototyping. The aim is to improve the capability of rapid product development, aiming at small and medium sized enterprises. Wang and Bi [29] presented the design of a new CAD/CAM teaching course incorporating both modern theory of manufacturing and CAD/CAM integrated tools. Integration of CAD with CAE systems is feasible as well. Louhichi et al. [30] proposed a method for the CAD model rebuilding after a FEA, by extracting geometric data from the analyzed mesh. Similar studies [31–33] seem to work on the integration of CAD with CAM or CAE systems in an effort to automate engineering tasks, fill gaps related to the interconnection between different types of software and to enhance the training process of the relevant personnel.

Benefits of Automated Design Based on CAD Programming. One of the primary benefits of automated design is the significant increase in efficiency. By automating repetitive and time-consuming tasks, designers can focus on more creative and strategic aspects of the design process. Scripting and parametric modeling allow for rapid iterations, enabling designers to explore more options in less time. This efficiency is particularly valuable in industries where time-to-market is a critical factor [3].

Automated design ensures a high level of precision and consistency in the final product. Algorithms can calculate exact dimensions, angles, and tolerances, reducing the risk of human error. This precision is essential in fields such as aerospace engineering, where even small deviations can lead to significant performance issues. Consistency is also crucial in mass production, where automated design ensures that every product meets the same quality standards [34].

Generative design, a form of CAD programming, enables innovation by exploring a vast design space that would be impractical to navigate manually. Designers can set performance criteria, such as weight, strength, and cost, and let the algorithm generate thousands of potential solutions. This approach often leads to novel designs that push the boundaries of traditional engineering and manufacturing [6].

Automated design can lead to significant cost savings by reducing the need for physical prototypes, minimizing material waste, and optimizing manufacturing processes. By simulating and testing designs in a virtual environment, companies can identify and address potential issues before they reach production. Moreover, automated design allows for easy customization and personalization of products, which is increasingly important in today's consumer-driven market. Parametric models can be adjusted to meet individual customer requirements, whether it's a custom-fit shoe or a tailored piece of furniture. This flexibility enhances customer satisfaction and opens up new business opportunities for companies that offer personalized products [35].

Challenges of Automated Design Based on CAD Programming. One of the main challenges of automated CAD programming is the complexity of the tools and techniques involved. Learning to write scripts, create parametric models, and develop algorithms requires a solid understanding of both programming and design principles. This learning curve can be steep, particularly for professionals who are more accustomed to traditional CAD methods. Moreover, as software tools continue to evolve, designers must keep up with new features and capabilities, requiring ongoing education and training.

Implementing automated design solutions can involve significant upfront costs, including software licenses, hardware, and training. High-end CAD software with advanced automation features is often expensive, and the cost of developing custom scripts or algorithms can add up quickly. While these investments often pay off in the long run through increased efficiency and reduced production costs, the initial financial barrier can be a challenge for small businesses or individual designers [9].

Integrating automated design tools into existing workflows can be challenging, particularly in companies with established processes and legacy systems. Ensuring compatibility between different software platforms, managing data interoperability, and aligning automated design with existing production methods requires careful planning and coordination. In some cases, companies may need to invest in additional software or develop custom solutions to bridge gaps between systems [36].

As with any digital process, automated design based on CAD programming raises concerns about data security and intellectual property protection. The use of cloud-based CAD platforms, in particular, exposes sensitive design data to potential cyber threats. Ensuring that *CAD data is securely stored and transmitted* is crucial, especially in industries where intellectual property is a key asset. According to Wang et al. [37] companies must implement robust security measures and establish clear policies for managing and protecting design data.

1.2.2 CAD-Based Programming for Manufacturing

CAD systems have revolutionized the manufacturing industry by offering tools that enable precise and efficient design and production processes. CAD-based programming, which involves using CAD models to drive manufacturing operations, has become integral to modern manufacturing. This integration has streamlined workflows, reduced errors, and enhanced the ability to produce complex geometries with high precision. The evolution of CAD-based programming aligns with advancements in manufacturing technologies such as Computer Numerical Control (CNC) machining, additive manufacturing, and robotics, all of which rely heavily on CAD data. CAD systems provide a digital representation of a product, which is crucial for various stages of the manufacturing process. These systems allow engineers to design complex parts and assemblies, simulate their behavior under different conditions, and optimize them for production. In manufacturing, CAD models serve as the basis for generating tool paths, controlling machines, and ensuring that the final product meets design specifications. This process is essential for translating a designer's intent into physical reality. By the 1970s and 1980s, CAD software had begun to incorporate features that allowed for direct integration with CNC machines, enabling the automatic generation of G-code, the language used to control CNC machinery. Today, CAD-based programming extends beyond CNC machining to encompass a wide range of manufacturing technologies, including additive manufacturing (3D printing), robotic automation, and inspection systems. Modern CAD software packages, such as SolidWorks™, Autodesk Inventor™, and Siemens NX™, include sophisticated tools for simulation, optimization, and direct machine control, making them indispensable in advanced manufacturing environments. A number of studies indicate the use of the aforementioned CAD systems and their programming interfaces for the development of tools related to design and modeling automation. Zbiciak et al. [38] utilized the programming interface of Siemens NX™ for the development of a solid model generative tool. The paper focuses on the modeling process of spur gears. Similarly, Tzotzis et al. [39] employed the API of Solid-Works™ to facilitate the engineering routine required for the development of a standardized mechanical system. Garcia-Hernandez et al. [40] proposed a method for controlling the geometric and kinematic variables required for the manufacturing of elliptical and oval gears using wire electro-discharge machining (WEDM), by utilizing the programming interface of Siemens NX™. Poniatowska [41] presented a methodology that can be applied off-line to CNC machining processes on free-form surfaces. The proposed methodology introduces corrections compensations for the CNC programs.

Boolean operations can be applied iteratively on the solid models, simulating this way the cutting process. On similar principals are based studies related to the simulation of other machining processes such as milling [42] and turning [43]. The results and findings that can be acquired by such studies, are usually related to the modeling, prediction and evaluation of the surface quality [42–45] of the manufactured product and the cutting forces [46–48] generated during the manufacturing process. Besides the aforementioned parameters, several other characteristics can be measured and evaluated with the aid of CAD-based resources, including tool deflection [49] and chip geometry [50].

Integration of CAD with CAM. CAM is the use of software and computer-controlled machinery to automate a manufacturing process. CAD and CAM systems are often integrated into a single environment, allowing for seamless transition from design to production. This integration is crucial for ensuring that the manufacturing process accurately follows the design specifications.

The typical CAD to CAM workflow begins with the creation of a 3D model in a CAD software. Once the design is finalized, the CAD model is imported into a CAM software, where tool paths are generated based on the geometry of the model. These tool paths dictate how the manufacturing equipment will move to shape the material, whether through cutting, milling, or additive processes. Finally, the CAM software generates the machine code (e.g., G-code) that directly controls the manufacturing equipment. This workflow highlights the importance of precision in CAD models, as any inaccuracies or errors in the design can propagate through the CAM process and result in defective products. Moreover, the integration of CAD and CAM allows for simulations of the manufacturing process to be performed before actual production begins, reducing the risk of errors and material waste. Favi et al. [51] proposed a method for the collection of design to manufacture rules, to facilitate the cooperation between design and manufacturing departments. For this purpose, the 3D models are analyzed and geometrical features are retrieved to create a knowledge-based system. Stănășel et al. [52] employed the solid model feature recognition capabilities of Siemens NX™, to generate CNC code for the manufacturing of components. Xu et al. [53] proposed a CAD, CAM and Computer-Aided Process Planning (CAPP) framework integration for the automatic generation of tool path. Similarly, Fountas et al. [54] examined the possibilities of integrating CAD with CAM and CAPP modules for facilitating the interconnection between design, process planning and manufacturing, since these are the pylons of the product lifecycle management. The different aspects of CAD integration with other modules focus on the process chain [55, 56], including hybrid and intelligent manufacturing [57–60]. It should be noted that a great number of studies are related to dentistry [61–63], focusing on the design, manufacturing, surgical planning and placement simulation of implants [64–66], meaning that CAD/CAM systems affect deeply the medical sector.

Advances in CAD-Based Programming Technologies. The continuous evolution of manufacturing technologies has driven advances in CAD-based programming. Below are some key developments that have shaped modern manufacturing.

Feature-based machining (FBM) is an advanced CAM technique that automates the process of defining machining operations based on the features of a CAD model. In traditional CAM workflows, machinists manually select the machining operations for each part of the model. In contrast, FBM automatically identifies features such as holes, pockets, and bosses in the CAD model and assigns the appropriate machining operations. This not only speeds up the programming process but also reduces the likelihood of human error. Relevant studies reveal the increasing use of the FBM method. Zhang et al. [67] focused on the recognition of 2½D model features, from the CNC program itself. The method is applicable to milling operations and enables the integration of CNC programs with the feature recognition method for identification purposes of the manufacturing process plan. Huang et al. [68] introduced a new multi-level structuralized model for manufacturing reuse. The principle relies on capturing the important model data such as the machining features, which describe the interactions among them. Abdulghafour and Hassan [69] developed a methodology for the automatic recognition of CAD features related to freeform surfaces. Feature-based recognition and topology identification is widely applied to most machining processes such as drilling [70, 71] and milling [67, 72, 73].

Parametric and associative modeling in CAD systems [74, 75] allow for the creation of models that can be easily modified and updated. In parametric modeling, dimensions and constraints are used to define the shape of a model, meaning that changes to these parameters will automatically update the model. Associative modeling goes a step further by maintaining relationships between different parts of an assembly, so that changes in one part automatically propagate to others. These capabilities are essential for CAD-based programming, as they enable quick revisions and ensure that all parts of a product remain consistent with the design intent.

Simulation tools within CAD and CAM environments allow manufacturers to model and analyze the manufacturing process before actual production begins. These tools can simulate various aspects of the process, including tool paths, material behavior, and machine movements, helping to identify potential issues early on. The concept of the digital twin, which involves creating a virtual replica of a physical product or process, has further enhanced the capabilities of CAD-based programming. Digital twins allow for the monitoring of manufacturing processes in real time, in addition to the optimization, leading to improved efficiency and quality [76–78].

Applications of CAD-Based Programming in Manufacturing. CAD-based programming is applied across various manufacturing sectors, from automotive and aerospace to consumer electronics and medical devices. Each application has its unique requirements, but all benefit from the precision and efficiency that CAD-based programming offers.

In the aerospace and defense industries, the complexity and high precision required in manufacturing parts make CAD-based programming indispensable. Components for aircraft and spacecraft often involve intricate geometries and tight tolerances, which are difficult to achieve without advanced CAD and CAM tools. Furthermore, the use of materials such as titanium and composite materials, which are challenging to machine, necessitates precise programming and simulation to avoid costly errors and rework.

Similarly, the automotive industry has widely adopted CAD-based programming to enhance the efficiency of vehicle production. CAD models are used to design everything from individual components to entire vehicle assemblies. These models are then used to program CNC machines, robotic arms, and other automated systems on the production line. The integration of CAD with other digital tools, such as Product Lifecycle Management (PLM) systems [79–81], ensures that design changes are accurately reflected throughout the manufacturing process, reducing lead times and costs.

Medical device manufacturing requires extreme precision, especially for devices that are implanted in the human body. CAD-based programming allows manufacturers to design and produce complex geometries, such as those required for orthopedic implants and surgical instruments, with high accuracy. Additionally, the ability to simulate the manufacturing process and perform virtual testing of devices before production helps ensure that they meet stringent regulatory standards.

Challenges and Future Directions. While CAD-based programming has significantly advanced manufacturing capabilities, it also presents several challenges that need to be addressed as the technology continues to evolve.

One of the primary challenges in CAD-based programming for manufacturing is the management and interoperability of data across different systems. As manufacturing processes become more complex, the amount of data generated, from CAD models to machine code, increases exponentially. Ensuring that this data remains consistent and accessible across different platforms and departments is crucial for maintaining efficiency and quality. Interoperability between different CAD and CAM systems is also a significant challenge, as manufacturers often use a mix of software tools that may not be fully compatible with each other [82, 83].

The adoption of CAD-based programming requires a workforce with specialized skills in CAD, CAM, and CNC programming. As these technologies continue to evolve, ongoing training and skill development are essential to keep up with new tools and methods. This is particularly important in industries such as aerospace and medical device manufacturing, where the complexity of products and processes demands a high level of expertise [84].

Finally, the integration of CAD-based programming with emerging technologies such as artificial intelligence (AI), machine learning, and the Internet of Things (IoT) presents both opportunities and challenges. AI and machine learning [85, 86] can enhance CAD and CAM systems by automating routine tasks, optimizing tool paths, and predicting potential issues before they arise. However, integrating these technologies into existing workflows requires significant investment in new software, hardware, and training.

1.2.3 Automating Engineering Tasks Using CAD-Based Programming

CAD systems were initially developed as digital drafting tools, allowing engineers to create 2D drawings and later 3D models with greater precision than traditional manual drafting. However, as the capabilities of CAD systems expanded, so did their application. Modern CAD systems are now integrated with analysis tools, simulation environments, and manufacturing processes, making them central to the entire product development lifecycle. The integration of programming into CAD systems, often referred to as CAD-based programming, allows engineers to automate tasks that would otherwise require significant manual effort. This automation can range from simple macros that replicate repetitive tasks to complex algorithms that optimize entire design workflows.

Drawing Automation and Documentation. The creation of technical drawings and documentation is a fundamental aspect of the engineering design process. These drawings serve as blueprints for manufacturing, assembly, and inspection, making accuracy and consistency crucial. Traditionally, generating these drawings from 3D models has been a labor-intensive task, prone to human error. CAD-based programming offers powerful tools to automate this process, significantly improving efficiency and reducing the potential for mistakes.

One major field of application is the automation of drawing creation by generating 2D technical drawings directly from 3D models. By writing scripts that interact with the CAD software's API, engineers can automate the generation of detailed drawings that include dimensions, annotations, part numbers, and material specifications. This process ensures that the drawings are always consistent with the 3D model, reducing the likelihood of discrepancies that can occur when drawings are manually updated. For example, in industries such as automotive or aerospace, where hundreds or even thousands of parts need to be documented, CAD-based programming can automatically generate these drawings based on predefined templates and rules. This not only saves time but also ensures that all drawings conform to industry standards and company-specific guidelines. Tzotzis et al. [39] employed the API of SolidWorks™ to develop a set of applications that deal with the automation of routine engineering tasks, including the automated mechanical drawing generation from solid parts. Similarly, Byun and Sohn [87] proposed a software system for automatically generating BIM models from two-dimensional building drawings. Most related studies focus on mechanical systems [88–90] and constructions [91, 92].

Another critical aspect of drawing automation is the automatic generation of annotations and dimensions [93]. CAD-based programming can be used to create scripts that apply consistent annotation styles across all drawings, ensuring that every detail is clearly communicated. Dimensions can be automatically generated based on the geometry of the 3D model, and these scripts can be customized to apply different dimensioning standards depending on the project requirements. Moreover, changes made to the 3D model can be automatically reflected in the 2D drawings, ensuring that all documentation is up-to-date without manual intervention. This is particularly important in iterative design processes, where changes are frequent, and maintaining accurate documentation is essential.

Automation in drawing generation significantly reduces the potential for human error. Manual creation of drawings is not only time-consuming but also susceptible to mistakes, such as missing dimensions, incorrect annotations, or inconsistent styling. By automating these tasks, engineers can ensure that every drawing is accurate and adheres to predefined standards, improving the overall quality of the design documentation. Furthermore, automated drawing generation allows for the rapid creation of updated drawings whenever changes are made to the 3D model, ensuring that all team members are working with the most current information.

Simulation and Analysis Automation. Simulation and analysis are critical components of the engineering design process, enabling engineers to evaluate the performance of their designs under various conditions. Traditionally, these tasks require significant manual effort, as engineers must set up simulations, define boundary conditions, and analyze results. CAD-based programming offers a way to automate these processes, allowing for faster and more thorough analysis.

One of the most significant benefits of CAD-based programming in simulation is the automation of the setup process. Scripts can be written to automatically define boundary conditions, apply loads, and select materials based on the geometry of the 3D model. This automation ensures that simulations are set up consistently and according to predefined criteria, reducing the potential for errors that can occur during manual setup. For example, in structural engineering, a script could automatically apply different load cases to a building model, run FEA, and compile the results into a report. This not only saves time but also allows engineers to explore a wider range of scenarios, ensuring that the design is robust under various conditions. Shaqura and Shamma [94] proposed a method for generating realistic models of an aerial drone mathematically and visually, by utilizing physics simulation and API resources. Gherardini et al. [74] developed a module for automatically analyzing the geometrical data of a pump 3D model. The acquired data can be directly used for fluid dynamic analysis. Ernst et al. [95] investigated the possibilities of CAD-based parametric model simulation tools for the automotive body design, focusing on the corrosion protection measures. The same principles apply to the robot operation simulation. Bedaka and Lin [96] developed a platform for generating a robot path from a CAD model and simulate the generated trajectories. Zheng et al. [97] utilized similar techniques and methodologies for generating the path of a laser cladding robot, specialized in the additive manufacturing industry. The path is automatically generating via a solid model and the simulation is performed according to the generated instructions.

CAD-based programming also enables the automation of batch processing, where multiple simulations are run in sequence. This is particularly useful for parametric studies, where the goal is to understand how changes in design parameters affect the performance of the model. By automating the simulation process, engineers can quickly evaluate numerous design alternatives, identifying the optimal configuration without manual intervention. For instance, in the design of an aircraft wing, a script could automatically vary parameters such as wing length, curvature, and material properties, running simulations for each configuration. The results can then be analyzed to determine the best design based on criteria such as weight, strength, and aerodynamics. Once simulations are completed, the results must be analyzed and documented. CAD-based programming can automate this process by generating reports that compile simulation results, including stress distributions, deformation plots, and safety factors. However, batch processing and automated result documentation are usually delt with macros and scripts. Zhang et al. [98] utilized the programming interface of SIEMENS NX™ to develop pre-processing tools for additive manufacturing. Lai et al. [99] utilized similar methods and techniques to integrate CAD tool with Abaqus™ FEA software for industrial applications. Wang et al. [100] developed an integrated framework for structural design optimization by using CAD and CAE software APIs, as well as scripting languages.

Another topic where CAD-based programming finds applicability in manufacturing is the assembly and disassembly of systems and complex products. Campi et al. [101] defined a generalized framework for coupling design for manufacturing and assembly rules with geometrical product features. By analyzing the 3D CAD model, any manufacturing issues can be anticipated and the cost can be controlled, during the development of the product. Chereshnia and Berman [102] proposed a two-stage algorithm, developed with SolidWorks™ API, for identifying the base component of an assembly, which can be assembled by standardized robotic arms. Similarly, Chervinskii et al. [103] proposed a framework for the automated assembly process, directly form CAD files, by utilizing a two robot cell. Assembly automation benefit from automated feature recognition from CAD models, expanding its use to additive manufacturing process optimization, sequence planning, and instructions generation [104–107].

Customization and Mass Customization. Customization has become increasingly important in industries ranging from consumer products to medical devices. Customers now expect products that meet their specific needs, whether it's a personalized smartphone case or a custom-fitted prosthetic limb. CAD-based programming plays a crucial role in enabling both individual customization and mass customization, where products are tailored to individual requirements on a large scale.

CAD-based programming allows for the automation of custom product design by creating parametric models that can be easily modified based on customer input. For example, in the design of custom eyewear, a script could adjust the size, shape, and style of the frames based on the customer's facial measurements and preferences. This process ensures that the final product fits the customer's specifications precisely. In the medical field, CAD-based programming can be used to design custom implants or prosthetics. By inputting patient-specific data, such as medical images, scripts can automatically generate 3D models that are perfectly tailored to the patient's anatomy. This level of customization ensures a better fit and improved functionality, leading to better patient outcomes.

Mass customization involves producing large quantities of customized products without sacrificing efficiency. CAD-based programming facilitates this by automating the design adjustments needed to accommodate individual customer requirements. For instance, in the automotive industry, customers can choose from a variety of options, such as paint colors, interior finishes, and wheel designs. CAD-based programming can automatically apply these choices to the base model, generating a customized design ready for production. An example of CAD-based automated application for customized design of standardized bicycles was presented by Kyratsis et al. [22] and another for a mechanical actuation system by Tzotzis et al. [39]. Moreover, CAD-based programming can be integrated with manufacturing processes to automate the generation of toolpaths, support structures, and other manufacturing-specific features directly from the CAD model, as already discussed. This integration ensures that customized designs can be produced efficiently, without the need for manual intervention at every step. One of the main advantages of CAD-based programming in customization and mass customization is the reduction in lead times and costs. By automating the design process, companies can quickly respond to customer orders, reducing the time from order to delivery. This speed is particularly important in industries like fashion or electronics, where customer preferences change rapidly, and companies need to stay ahead of trends. Yuan and Huh [108] approached the mass production system for customized apparel, by enabling communication between the CAD system and other related modules. The authors employed the advantages of CAD systems as well as the benefits of programming. Similarly, Manavis et al. [109] utilized Grasshopper™ visual programming language to develop a design tool for the furniture industry, generating customized models under a wide variety of editable parameters. Binder et al. [110] investigated the development of a framework for the mass customization of wearable heat exchangers used in the healthcare industry. Gembarski and Gast [111] presented an approach of parametric and knowledge-based modeling with CAD systems to develop an interactive configuration system for piping. Earlier studies [23, 39, 108] shown that mass customization is feasible by implementing CAD programming to a generalized framework for design and manufacturing of standardized products and systems, as well as apparel. Furthermore, automation reduces the cost of customization by minimizing the amount of manual labor required. Once a script is developed to automate a particular customization process, it can be reused for multiple orders, spreading the development cost over a large number of units. This scalability makes mass customization economically viable, even for complex products.

1.3   Research Summary

The research papers used in this review were selected from within the period between 2014 and 2024. Three main research areas: design modeling and optimization, manufacturing and assembly, as well as simulation and analysis, constitute the pool from which the papers were drawn. The total number of referenced works is 111. However, four books and one paper from the 1960's that were used to supplement the review, were not taken into account. Thus, the number of the papers that are displayed in this section is 106.

Figure 1.1 visualizes the distribution of the papers per year. It is evident that the 42.45% of the total papers were drawn from the period between 2014 and 2019, whereas the rest 57.55% from the period between 2020 and 2024.

Figure 1.2 illustrates the distribution of the classification of the papers per main research area. It is shown that the majority of the papers (76.4% of total papers) belongs to either the 1st area (design, modeling and optimization) or the 2nd one

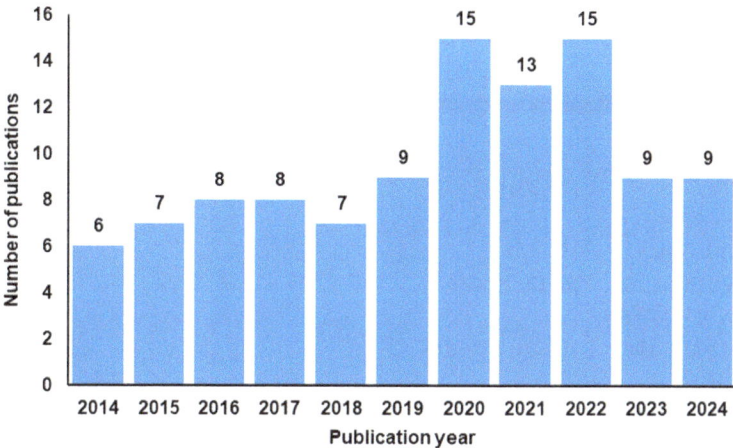

Fig. 1.1 Distribution of publications per year

(manufacturing and assembly), with a percentage of 38.7% and 37.7% respectively. The rest 23.6% belongs to the 3rd area (simulation and analysis) and the combinations between the three areas. This means that some papers are included in two main research areas. It should be noted that no paper is included to all three areas.

To capture the essence of the three main research areas, they were divided into groups with respect to the research topic the collected papers deal with. Therefore, a classification was made according to the title, the keywords and the content of each paper. It should be noted that some papers were classified to more than one group. For the design, modeling and optimization research area, the next groups were defined, as shown in Table 1.1:

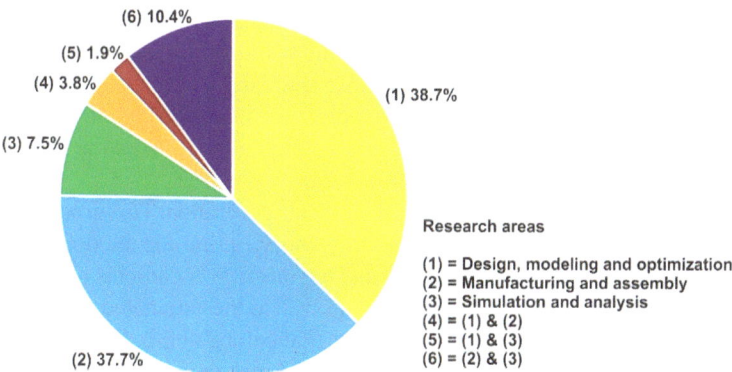

Fig. 1.2 Classification of papers per research area

Table 1.1 Classification of the 1st research area's papers per group

Group	Papers
Automated design and modeling, parametric design	[2, 4, 5, 7, 8, 11–13, 17, 19–23, 36, 38, 39, 74, 75, 87–91, 93, 97, 110]
Design for manufacturing	[27, 33, 51]
Generative design	[14, 15]
Topology, shape and structure optimization	[6, 18, 25, 99]
Systems and products customization	[11, 34, 107–109]
Automated drawing and engineering documents generation	[39, 82, 86, 91, 92]
CAD and PLM integration	[79–81]

- Automated design and modeling, parametric design. This group constitutes the majority of the collected papers of the first area and deals with CAD-based programming, scripting and similar programming methodologies for the design or modeling automation of products and systems. Additionally, some papers in this group relate to parametric design.
- Design for manufacturing. This group deals with techniques used during the development stage of a product, that allow its efficient and convenient manufacturing. Cost, time, and resources are the main factors considered.
- Generative design. These papers highlight the significance of generative design, benefits, as well as applications.
- Topology, shape, and structure optimization. This group focuses on the techniques, strategies, and methodologies utilized for the optimization of models, in terms of shape, weight, strength, and other critical parameters.
- Systems and products customization. Papers of this group typically target the automated generation of product design configurations, by taking into account the customer preferences.
- Automated drawing and engineering documents generation. This group highlights the potential of CAD-based programming and scripting for the development of tools used for the automated drawing and engineering documents generation, the automated dimensioning and annotation, as well as the automated documentation.
- CAD and PLM integration. Finally, papers that belong to this group deal with the integration of CAD systems with PLM for the realization of more efficient and manageable product data systems.

For the manufacturing and assembly research area, the next groups were defined, as shown in Table 1.2:

Table 1.2 Classification of the 2nd research area's papers per group

Group	Papers
Automated assembly	[12, 102]
CAD and CAM integration	[28, 40, 46, 49, 54–58, 61–66, 69, 83]
CAD/CAM and CAE integration	[31, 47, 49]
Design for assembly and part preparation	[16, 33, 93, 100, 101, 103, 105]
Smart manufacturing and digital twins	[37, 59, 60, 76–78, 81]
Feature recognition for manufacturing	[52, 67–71, 84, 85, 104, 106]
Machining analysis and process parameters measurement	[41–46, 48, 50, 71–73]
CNC toolpath generation	[26, 53]

- Automated assembly. This group regards the use of programming resources for the automatic realization of large mechanical assemblies.
- CAD and CAM integration. This group includes the most of the collected papers from the second research area. The studies focus mostly on the integration of CAD and CAM systems for the optimization of CNC programs, as well as the automation of certain tasks related to the CNC programming.
- CAD/CAM and CAE integration. Similarly to the previous group, this one demonstrates the integration of all major pieces of software, including simulation modules as well. The purpose of this integration is to further enhance the CNC programming processes and manufacturing procedures.
- Design for assembly and part preparation. Design techniques and methodologies that are used to facilitate the assembly process are the content of this group. The concept of these studies rely on the definition of rules that allow the preparation of a part in such a way that it can be conveniently connected to other parts in order to form an assembly.
- Smart manufacturing and digital twins. These papers focus on the CAD systems implementation in the digital twin processing and smart manufacturing.
- Feature recognition for manufacturing. This group is closely related to the CAD and CAM integration, and focuses on design and manufacturing features recognition techniques and methodologies. Feature recognition is an indispensable aspect of CAD-based manufacturing.
- Machining analysis and process parameters measurement. This group deals with the analysis of machining processes, the CAD-based simulation of standard machining and manufacturing processes, as well as the measurement of typical parameters such as cutting forces, surface roughness, chip geometry, and surface topography.
- CNC toolpath generation. Here, the papers focus on the automated design feature recognition and manufacturing features extraction for the optimization of the process planning or CNC machine toolpath.

Table 1.3 includes the groups for the simulation and analysis research area:

Table 1.3 Classification of the 3rd research area's papers per subcategories

Subcategory	Papers
CAD and CAE integration	[30–32, 49, 98, 99]
Systems and processes simulation	[24, 42–45, 48, 50, 72–74, 94–96]
Tools for FEA	[29, 47, 98]

- CAD and CAE integration. This group includes papers that deal solely with the CAD and CAE software integration for facilitating simulation analyses and data management directly from the CAD environment or the interconnection between CAD systems and CAE software.
- Systems and processes simulation. These papers presents the development of CAD-based simulations of machining processes and manufacturing procedures, as well as the simulation of robotic and similar systems.
- Tools for FEA. These papers refer to the development of tools that are related to the assistance of FEA in terms of efficient and automated mesh generation and general setup, as well as the processing of manufacturing processes such as additive manufacturing.

1.4   Future Trends and Directions

The integration of AI and machine learning with CAD programming is poised to revolutionize automated design. AI algorithms can analyze vast amounts of design data to identify patterns and make predictions, helping designers optimize their models in ways that were previously impossible. For example, AI-driven generative design can explore millions of potential solutions, learning from each iteration to improve the final outcome. This approach could lead to more efficient, innovative, and sustainable designs across various industries.

Cloud-based CAD platforms are increasingly enabling real-time collaboration between designers, engineers, and other scientists. These platforms allow multiple users to work on the same model simultaneously, regardless of their physical location. Automated design tools are being integrated into these cloud environments, making it easier for teams to develop, test, and optimize designs collaboratively. This trend is particularly relevant in global industries, where teams are often distributed across different regions.

The rise of additive manufacturing is closely linked with advancements in automated CAD programming. As 3D printing technology evolves, CAD software is being developed to automatically generate models that are optimized for additive manufacturing. This includes considerations for material usage, print time, and structural integrity. Automated design tools can also create complex geometries that would be difficult or impossible to produce using traditional manufacturing methods, further expanding the possibilities of additive manufacturing.

As sustainability becomes a priority in design and manufacturing, automated CAD programming will play a key role in developing environmentally friendly solutions. Generative design algorithms can optimize material usage, minimize waste, and reduce energy consumption, leading to more sustainable products and processes. Additionally, CAD software can be used to simulate and analyze the environmental impact of different design choices, helping companies make more informed decisions about materials, production methods, and product lifecycles.

1.5 Conclusion

The present review focuses on the collection of research works and studies related to the implementation of CAD-based programming, scripting and similar resources in the design and manufacturing of products and systems. The collected papers were drawn from the period 2014–2024 and were allocated to three main research areas: design, modeling and optimization; manufacturing and assembly; simulation and analysis. In addition, each research area was divided into a number of groups depending on the content of the papers it included. By realizing the classification of the papers, an analysis of their content was performed, revealing the benefits, challenges, applications and future trends of CAD programming in design and manufacturing. Summarizing, the next conclusions can be drawn.

Automated design based on CAD programming represents a significant advancement in the fields of design, engineering, and manufacturing. By leveraging scripting, parametric modeling, and algorithmic design, professionals can automate complex tasks, optimize designs, and create innovative solutions that push the boundaries of traditional methods. While there are challenges to overcome, such as the complexity of the tools and the high initial costs, the benefits of increased efficiency, precision, and customization are substantial. As technology continues to evolve, the integration of AI, cloud computing, and additive manufacturing with CAD programming will further enhance the capabilities of automated design, leading to a future where products are designed faster, smarter, and more sustainably than ever before.

Moreover, CAD-based programming can be considered a cornerstone of modern manufacturing, enabling the precise and efficient production of complex parts and assemblies. The integration of CAD with CAM and other digital tools has transformed the way manufacturers design, simulate, and produce products. As manufacturing technologies continue to advance, CAD-based programming will play an increasingly important role in driving innovation and competitiveness in the industry. However, to fully realize the potential of CAD-based programming, manufacturers must address challenges related to data management, workforce training, and the integration of emerging technologies. By doing so, they can continue to leverage CAD-based programming to enhance productivity, reduce costs, and improve the quality of their products.

Finally, CAD-based programming in general represents a significant advancement in the field of engineering, offering powerful tools for automating design tasks, optimizing processes, and enhancing productivity. By integrating programming with CAD systems, engineers can automate repetitive tasks, explore a wider range of design alternatives, and ensure consistency and accuracy throughout the design process. However, the adoption of CAD-based programming requires overcoming challenges related to skill development, software integration, and maintenance. As new technologies like AI, cloud computing, and advanced manufacturing continue to evolve, the potential for CAD-based programming will only grow, paving the way for more efficient, innovative, and intelligent engineering solutions.

References

1. Sutherland IE (1964) Sketch pad a man-machine graphical communication system. Proc Des Autom Conf 23.https://doi.org/10.1145/800265.810742
2. Camba JD, Contero M, Company P (2016) Parametric CAD modeling: an analysis of strategies for design reusability. Comput Des 74:18–31. https://doi.org/10.1016/j.cad.2016.01.003
3. McMahon C, Browne J (1999) CADCAM: principles, practice and manufacturing management, 2nd edn. Addison-Wesley Longman Publishing Co., Inc, USA
4. Moreno R, Bazán AM (2017) Design automation using script languages. High-level CAD templates in non-parametric programs. IOP Conf Ser Mater Sci Eng 245. https://doi.org/10.1088/1757-899X/245/6/062039
5. Marschallinger R, Jandrisevits C, Zobl F (2015) A visual LISP program for voxelizing AutoCAD solid models. Comput Geosci 74:110–120. https://doi.org/10.1016/j.cageo.2014.09.011
6. Panesar A, Abdi M, Hickman D, Ashcroft I (2018) Strategies for functionally graded lattice structures derived using topology optimisation for additive manufacturing. Addit Manuf 19:81–94. https://doi.org/10.1016/j.addma.2017.11.008
7. Gella-Marín R, Tzotzis A, García-Hernández C, et al (2021) CAD software integration with programming tools for modelling, measurement and verification of surfaces. In: Experiments and simulations in advanced manufacturing, pp 91–116
8. Tzotzis A, García-Hernández C, Huertas-Talón JL, Kyratsis P (2020) CAD-based automated design of FEA-ready cutting tools. J Manuf Mater Process 4:1–14. https://doi.org/10.3390/jmmp4040104
9. Groover MP (2016) Automation, production systems, and computer-integrated manufacturing. Pearson Education India
10. Pottmann H, Asperl A, Kililan A (2007) Architectural geometry. SIAM
11. Manavis A, Tzotzis A, Tsagaris A, Kyratsis P (2022) A novel computational-based visual brand identity (CbVBI). Machines
12. Kyratsis P, Tzotzis A, Manavis A (2021) Computational design and digital fabrication. In: Kumar S, Rajurkar KP (eds) Advances in manufacturing systems. Springer Singapore, Singapore, pp 1–16
13. Kopylov M (2021) Expanding the functionality of optical CAD using the Python scripting language. CEUR Workshop Proc 3027:172–179. https://doi.org/10.20948/graphicon-2021-3027-172-179
14. Ntintakis I, Stavroulakis GE (2020) Progress and recent trends in generative design. MATEC Web Conf 318:01006. https://doi.org/10.1051/matecconf/202031801006
15. Junk S, Burkart L (2021) Comparison of CAD systems for generative design for use with additive manufacturing. Procedia CIRP 100:577–582. https://doi.org/10.1016/j.procir.2021.05.126

16. Bacciaglia A, Ceruti A, Livcrani A (2022) Towards large parts manufacturing in additive technologies for aerospace and automotive applications. Procedia Comput Sci 200:1113–1124. https://doi.org/10.1016/j.procs.2022.01.311
17. Forrai M, Gavačová J, Gulan L (2016) A practical methodology for creating robust parametric surface-based models in automotive engineering. Procedia CIRP 50:484–489. https://doi.org/10.1016/j.procir.2016.04.145
18. Agarwal D, Marques S, Robinson TT (2022) Aerodynamic shape optimisation using parametric CAD and discrete adjoint. Aerospace 9:1–17. https://doi.org/10.3390/aerospace9120743
19. Hirz M, Rossbacher P, Gulanová J (2017) Future trends in CAD—from the perspective of automotive industry. Comput Aided Des Appl 14:734–741. https://doi.org/10.1080/16864360.2017.1287675
20. Wiberg A, Persson JA, Ölvander J (2023) A design automation framework supporting design for additive manufacturing
21. Dalpadulo E, Pini F, Leali F (2020) Integrated CAD platform approach for design for additive manufacturing of high performance automotive components. Int J Interact Des Manuf 14:899–909. https://doi.org/10.1007/s12008-020-00684-7
22. Kyratsis P, Gabis E, Tzotzis A et al (2019) CAD based product design: a case study. Int J Mod Manuf Technol 11:88–93
23. Kyratsis P, Tzotzis A, Tzetzis D, Sapidis N (2018) Pneumatic cylinder design using cad-based programming. Acad J Manuf Eng 16:107–113
24. Tsagaris A, Polychroniadis C, Tzotzis A, Kyratsis P (2024) Cost-effective robotic arm simulation and system verification. Int J Intell Syst Appl 16:1–12. https://doi.org/10.5815/ijisa.2024.02.01
25. Li M, Lin C, Chen W et al (2023) XVoxel-based parametric design optimization of feature models. Comput Des 160:103528. https://doi.org/10.1016/j.cad.2023.103528
26. Kukreja A, Dhanda M, Pande SS (2020) Efficient toolpath planning for voxel-based cnc rough machining. Comput Aided Des Appl 18:285–296. https://doi.org/10.14733/cadaps.2021.285-296
27. Favi C, Campi F (2021) CAD-based design for welding (DFW) method. Int J Interact Des Manuf 15:95–97. https://doi.org/10.1007/s12008-020-00727-z
28. Matta AK, Raju DR, Suman KNS (2015) The integration of CAD/CAM and rapid prototyping in product development: a review. Mater Today Proc 2:3438–3445. https://doi.org/10.1016/j.matpr.2015.07.319
29. Wang X, Bi Z (2019) New CAD/CAM course framework in digital manufacturing. Computer Applications in Engineering Education 27(1) 128-144 . https://doi.org/10.1002/cae.v27.1. https://doi.org/10.1002/cae.22063
30. Louhichi B, Abenhaim GN, Tahan AS (2015) CAD/CAE integration: updating the CAD model after a FEM analysis. Int J Adv Manuf Technol 76:391–400. https://doi.org/10.1007/s00170-014-6248-y
31. Fernandes FAO, Fuchter Júnior N, Daleffe A et al (2020) Integrating CAD/CAE/CAM in engineering curricula: a project-based learning approach. Educ Sci 10.https://doi.org/10.3390/educsci10050125
32. Kirkwood R, Sherwood JA (2018) Sustained CAD/CAE integration: integrating with successive versions of step or IGES files. Eng Comput 34:1–13. https://doi.org/10.1007/s00366-017-0516-z
33. Claudio FC, Favi MG, Mandolini M (2022) CAD-integrated design for manufacturing and assembly in mechanical design. Int J Comput Integr Manuf 35:282–325.https://doi.org/10.1080/0951192X.2021.1992659
34. Zawadzki P, Zywicki K (2016) Smart product design and production control for effective mass customization in the industry 4.0 concept. Manag Prod Eng Rev 7:105–112. https://doi.org/10.1515/mper-2016-0030
35. Piller FT (2010) Handbook of research in mass customization and personalization. World scientific

36. Salchner M, Stadler S, Hirz M et al (2016) Multi-CAD approach for knowledge-based design methods. Comput Aided Des Appl 13:471–483. https://doi.org/10.1080/16864360.2015.113 1540
37. Wang L, Törngren M, Onori M (2015) Current status and advancement of cyber-physical systems in manufacturing. J Manuf Syst 37:517–527. https://doi.org/10.1016/j.jmsy.2015. 04.008
38. Zbiciak M, Grabowik C, Janik W (2015) An automation of design and modelling tasks in NX Siemens environment with original software—generator module. IOP Conf Ser Mater Sci Eng 95.https://doi.org/10.1088/1757-899X/95/1/012117
39. Tzotzis A, Garcia-Hernandez C, Huertas-Talon J-L, et al (2017) Engineering applications using CAD based application programming interface. In: MATEC web of conferences, pp 1–7
40. Garcia-Hernandez C, Marín R, Talón J et al (2016) WEDM manufacturing method for noncircular gears using CAD/CAM software. Strojniški Vestn J Mech Eng 62:137–144. https://doi. org/10.5545/sv-jme.2015.2994
41. Poniatowska M (2015) Free-form surface machining error compensation applying 3D CAD machining pattern model. Comput Des 62:227–235. https://doi.org/10.1016/j.cad.2014. 12.003
42. Tapoglou N, Efstathiou C, Tzotzis A, Kyratsis P (2023) Study of the topography of face milled surfaces using CAD-based simulation. In: Kyratsis P, Manavis A, Davim JP (eds) Computational design and digital manufacturing. Springer International Publishing, Cham, pp 159–166
43. Tzotzis A, Tsagaris A, Tapoglou N, Kyratsis P (2023) High-precision CAD-based simulation for turning considering tool microgeometry. Int J Mechatron Manuf Syst 16:83–95. https:// doi.org/10.1504/IJMMS.2023.132023
44. Felhő C, Varga G (2022) CAD and FEM modelling of theoretical roughness in diamond burnishing. Int J Precis Eng Manuf 23:375–384. https://doi.org/10.1007/s12541-022-00622-5
45. Vakondios DG, Kyratsis P (2020) An innovative CAD—based simulation of ball—end milling in microscale. Adv Comput Des 5:13–34. https://doi.org/10.12989/acd.2020.5.1.013
46. Käsemodel RB, de Souza AF, Voigt R et al (2020) CAD/CAM interfaced algorithm reduces cutting force, roughness, and machining time in free-form milling. Int J Adv Manuf Technol 107:1883–1900. https://doi.org/10.1007/s00170-020-05143-x
47. Tzotzis A, García-Hernández C, Huertas-Talón J-L, Kyratsis P (2020) Influence of the nose radius on the machining forces induced during AISI-4140 hard turning: a CAD-based and 3D FEM approach. Micromachines 11:798. https://doi.org/10.3390/mi11090798
48. Tapoglou N (2019) Calculation of non-deformed chip and gear geometry in power skiving using a CAD-based simulation. Int J Adv Manuf Technol 100:1779–1785. https://doi.org/10. 1007/s00170-018-2790-3
49. Wang L, Chen ZC (2014) A new CAD/CAM/CAE integration approach to predicting tool deflection of end mills. Int J Adv Manuf Technol 72:1677–1686. https://doi.org/10.1007/s00 170-014-5760-4
50. Marinakis A, Dandouti E, Antoniadis A (2022) CAD-Based Simulation model for the calculation of chip geometry and cutting force components in gear shaping. J Manuf Sci Eng. https://doi.org/10.1115/1.4053809
51. Favi C, Mandolini M, Campi F, Germani M (2021) A CAD-based design for manufacturing method for casted components. Procedia CIRP 100:235–240. https://doi.org/10.1016/j.procir. 2021.05.061
52. Stănășel I, Blaga FS, Buidoș T (2019) Manufacturing based on feature recognition using NX. IOP Conf Ser Mater Sci Eng 568.https://doi.org/10.1088/1757-899X/568/1/012015
53. Xu T, Chen Z, Li J, Yan X (2015) Automatic tool path generation from structuralized machining process integrated with CAD/CAPP/CAM system. Int J Adv Manuf Technol 80:1097–1111. https://doi.org/10.1007/s00170-015-7067-5
54. Fountas NA, Krimpenis AA, Vaxevanidis NM (2014) Software development tools to automate CAD/CAM systems. Smart Manuf Innov Transform Interconnect Intell 190–224.https://doi. org/10.4018/978-1-4666-5836-3.ch008

55. Manoharan T, Humpa M, Martha A, Koehler P (2016) Knowledge integration in CAD-CAM process chain. Comput Aided Des Appl 13:729–736. https://doi.org/10.1080/16864360.2016.1150720
56. Schmid J, Pichler R (2020) Seamless data integration in the CAM-NC process chain in a learning factory. Procedia Manuf 45:31–36. https://doi.org/10.1016/j.promfg.2020.04.038
57. Caligiana G, Francia D, Liverani A (2017) In: Eynard B, Nigrelli V, Oliveri SM et al (eds) CAD-CAM integration for 3D hybrid manufacturing BT—advances on mechanics, design engineering and manufacturing: proceedings of the international joint conference on mechanics, design engineering & advanced manufacturing (JCM 2016), 14–16 September, 20. Springer International Publishing, Cham, pp 329–337
58. Wang X, Bi Z (2019) New CAD/CAM course framework in digital manufacturing. Comput Appl Eng Educ 27:128–144. https://doi.org/10.1002/cae.22063
59. Mourtzis D, Angelopoulos J, Panopoulos N (2022) In: MacCarthy BL, Ivanov DBT-TDSC (eds) Chapter 2—digital manufacturing: the evolution of traditional manufacturing toward an automated and interoperable smart manufacturing ecosystem. Elsevier, pp 27–45
60. Besharati-Foumani H, Lohtander M, Varis J (2019) Intelligent process planning for smart manufacturing systems: a state-of-the-art review. Procedia Manuf 38:156–162. https://doi.org/10.1016/j.promfg.2020.01.021
61. Thalji G, Jia-mahasap W (2017) CAD/CAM removable dental prostheses: a review of digital impression techniques for edentulous arches and advancements on design and manufacturing systems. Curr Oral Heal Reports 4:151–157. https://doi.org/10.1007/s40496-017-0137-z
62. Rexhepi I, Santilli M, D'Addazio G, et al (2023) Clinical applications and mechanical properties of CAD-CAM materials in restorative and prosthetic dentistry: a systematic review. J Funct Biomater 14.https://doi.org/10.3390/jfb14080431
63. Solís Pinargote NW, Yanushevich O, Krikheli N, et al (2024) Materials and methods for all-ceramic dental restorations using computer-aided design (CAD) and computer-aided manufacturing (CAM) technologies—a brief review. Dent J 12.https://doi.org/10.3390/dj12030047
64. Barone S, Casinelli M, Frascaria M et al (2016) Interactive design of dental implant placements through CAD-CAM technologies: from 3D imaging to additive manufacturing. Int J Interact Des Manuf 10:105–117. https://doi.org/10.1007/s12008-014-0229-0
65. Cuéllar CN, Rial MT, Antúnez-Conde R et al (2021) Virtual surgical planning, stereolitographic models and cad/cam titanium mesh for three-dimensional reconstruction of fibula flap with iliac crest graft and dental implants. J Clin Med. https://doi.org/10.3390/jcm10091922
66. Ochandiano S, García-Mato D, Gonzalez-Alvarez A et al (2022) Computer-assisted dental implant placement following free flap reconstruction: virtual planning, CAD/CAM templates, dynamic navigation and augmented reality. Front Oncol 11:1–16. https://doi.org/10.3389/fonc.2021.754943
67. Zhang X, Nassehi A, Newman ST (2014) Feature recognition from CNC part programs for milling operations. Int J Adv Manuf Technol 70:397–412. https://doi.org/10.1007/s00170-013-5275-4
68. Huang R, Zhang S, Bai X, Xu C (2014) Multi-level structuralized model-based definition model based on machining features for manufacturing reuse of mechanical parts. Int J Adv Manuf Technol 75:1035–1048. https://doi.org/10.1007/s00170-014-6183-y
69. Abdulghafour AB, Hassan AT (2019) Feature recognition of freeform surfaces for CAD/CAM integration. In: 2019 1st international conference on electrical, control and instrumentation engineering (ICECIE), pp 1–6
70. Ghadai S, Balu A, Sarkar S, Krishnamurthy A (2018) Learning localized features in 3D CAD models for manufacturability analysis of drilled holes. Comput Aided Geom Des 62:263–275. https://doi.org/10.1016/j.cagd.2018.03.024
71. Tzotzis A, Manavis A, Efkolidis N, Kyratsis P (2021) CAD-based automated G-code generation for drilling operations. Int J Mod Manuf Technol 13:177–184. https://doi.org/10.54684/ijmmt.2021.13.3.177

72. Marin F, Fagali de Souza A, da Silva Gaspar H, et al (2024) Topography simulation of free-form surface ball-end milling through partial discretization of linearised toolpaths. Eng Sci Technol Int J 55.https://doi.org/10.1016/j.jestch.2024.101757

73. Denkena B, Dittrich MA, Huuk J (2021) Simulation-based surface roughness modelling in end milling. Procedia CIRP 99:151–156. https://doi.org/10.1016/j.procir.2021.03.096

74. Gherardini F, Zardin B, Leali F (2016) A parametric CAD-based method for modelling and simulation of positive displacement machines. J Mech Sci Technol 30:3253–3263. https://doi.org/10.1007/s12206-016-0634-3

75. Reddy EJ, Reddy KS, Reddy PN (2024) Recent advancements in knowledge-based parametric modeling of mechanical components. In: Yadav S, Arora PK, Sharma AK, Kumar H (eds) Proceedings of third international conference in mechanical and energy technology. Springer Nature Singapore, Singapore, pp 401–411

76. Hänel A, Schnellhardt T, Wenkler E et al (2020) The development of a digital twin for machining processes for the application in aerospace industry. Procedia CIRP 93:1399–1404. https://doi.org/10.1016/j.procir.2020.04.017

77. Roy RB, Mishra D, Pal SK et al (2020) Digital twin: current scenario and a case study on a manufacturing process. Int J Adv Manuf Technol 107:3691–3714. https://doi.org/10.1007/s00170-020-05306-w

78. Wu Q, Mao Y, Chen J, Wang C (2021) Application research of digital twin-driven ship intelligent manufacturing system: pipe machining production line. J Mar Sci Eng 9.https://doi.org/10.3390/jmse9030338

79. Nzetchou S, Durupt A, Remy S, Eynard B (2022) Semantic enrichment approach for low-level CAD models managed in PLM context: literature review and research prospect. Comput Ind 135:103575. https://doi.org/10.1016/j.compind.2021.103575

80. Guyon S, Legoubé L-A, Terrier P, Rivest L (2024) In: Stark J (ed) The role of CAD and PLM in ecodesign: a short review BT—product lifecycle management (volume 6): increasing the value of PLM with innovative new technologies. Springer Nature Switzerland, Cham, pp 43–77

81. Gwangwava N (2018) Open product lifecycle management (PLM) for cloud manufacturing and cloud-based maintenance integration using XML. Adv E-bus Res Ser. https://doi.org/10.4018/978-1-5225-3628-4.ch011

82. Ramnath S, Haghighi P, Venkiteswaran A, Shah JJ (2020) Interoperability of CAD geometry and product manufacturing information for computer integrated manufacturing. Int J Comput Integr Manuf 33:116–132. https://doi.org/10.1080/0951192X.2020.1718760

83. Idrissi Gartoumi K, Zaki S, Aboussaleh M (2023) Building information modelling (BIM) interoperability for architecture and engineering (AE) of the structural project: a case study. Mater Today Proc. https://doi.org/10.1016/j.matpr.2023.05.408

84. Ikubanni PP, Adeleke AA, Agboola OO et al (2022) Present and future impacts of computer-aided design/computer-aided manufacturing (CAD/CAM). J Eur des Syst Autom 55:349–357. https://doi.org/10.18280/jesa.550307

85. Patel S, Mekavibul J, Park J, et al (2019) Using machine learning to analyze image data from advanced manufacturing processes. In: 2019 systems and information engineering design symposium (SIEDS), pp 1–5

86. Peddireddy D, Fu X, Wang H et al (2020) Deep learning based approach for identifying conventional machining processes from CAD data. Procedia Manuf 48:915–925. https://doi.org/10.1016/j.promfg.2020.05.130

87. Byun Y, Sohn BS (2020) ABGS: a system for the automatic generation of building information models from two-dimensional CAD drawings. Sustain 12.https://doi.org/10.3390/SU12176713

88. Esanakula JR, Raju GN, Sai GA (2020) Development of a preliminary system for automatic generation of CAD model of the universal coupling. Int J Mech Prod Eng Res Dev 10:8701–8710

89. Willis KDD, Jayaraman K, Lambourne JG et al (2021) Engineering sketch generation for computer-aided design engineering sketch generator sketch generation 3D synthesis 3D composition. Proc IEEE/CVF Conf Comput Vis Pattern Recognit

90. Long X, Li H, Du Y et al (2021) A knowledge-based automated design system for mechanical products based on a general knowledge framework. Expert Syst Appl 178:114960. https://doi.org/10.1016/j.eswa.2021.114960
91. Goldbach AK, Lázaro C (2024) CAD-integrated parametric design and analysis of lightweight shell structures. Structures 64:106566. https://doi.org/10.1016/j.istruc.2024.106566
92. Jin S, Zhang Y, Yamazaki T, Jiang Z (2021) Automatic 3D CAD model and 2D drawings generation in construction engineering. J Phys Conf Ser. https://doi.org/10.1088/1742-6596/1827/1/012115
93. Kakoulis KG, Konstantinidis P, Manavis A, Kyratsis P (2023) A system for automatic dimensioning of mechanical drawings. Int J Mod Manuf Technol 15:67–74. https://doi.org/10.54684/ijmmt.2023.15.3.67
94. Shaqura M, Shamma JS (2017) An automated quadcopter CAD based design and modeling platform using solidworks API and smart dynamic assembly. ICINCO 2017—Proc 14th Int Conf Informatics Control Autom Robot 2:122–131. https://doi.org/10.5220/0006438601220131
95. Ernst M, Hirz M, Stadler S (2014) A method of CAD based automation and simulation by the example of virtual stone chipping testing. Comput Aided Des Appl 11:295–304. https://doi.org/10.1080/16864360.2014.863495
96. Bedaka AK, Lin CY (2018) CAD-based robot path planning and simulation using OPEN CASCADE. Procedia Comput Sci 133:779–785. https://doi.org/10.1016/j.procs.2018.07.119
97. Zheng H, Cong M, Dong H et al (2017) CAD-based automatic path generation and optimization for laser cladding robot in additive manufacturing. Int J Adv Manuf Technol 92:3605–3614. https://doi.org/10.1007/s00170-017-0384-0
98. Zhang B, Goel A, Ghalsasi O, Anand S (2019) CAD-based design and pre-processing tools for additive manufacturing. J Manuf Syst 52:227–241. https://doi.org/10.1016/j.jmsy.2019.03.005
99. Lai Y, Zhang YJ, Liu L et al (2017) Integrating CAD with Abaqus: a practical isogeometric analysis software platform for industrial applications. Comput Math Appl 74:1648–1660. https://doi.org/10.1016/j.camwa.2017.03.032
100. Wang D, Hu F, Ma Z et al (2014) A CAD/CAE integrated framework for structural design optimization using sequential approximation optimization. Adv Eng Softw 76:56–68. https://doi.org/10.1016/j.advengsoft.2014.05.007
101. Campi F, Favi C, Germani M, Mandolini M (2022) CAD-integrated design for manufacturing and assembly in mechanical design. Int J Comput Integr Manuf 35:282–325. https://doi.org/10.1080/0951192X.2021.1992659
102. Chereshnia S, Berman S (2022) Automatic identification of the assembly base component for robotic manufacturing. IFAC-PapersOnLine 55:90–95. https://doi.org/10.1016/j.ifacol.2022.04.175
103. Chervinskii F, Zobov S, Rybnikov A et al (2023) Auto-assembly: a framework for automated robotic assembly directly from CAD. In: 2023 IEEE international conference on robotics and automation (ICRA), pp 11294–11300
104. Münker S, Schmitt RH (2022) CAD-based AND/OR graph generation algorithms in (dis)assembly sequence planning of complex products. Procedia CIRP 106:144–149. https://doi.org/10.1016/j.procir.2022.02.169
105. Gonnermann C, Gebauer D, Daub R (2023) CAD-based feature recognition for process monitoring planning in assembly. Appl Sci 13.https://doi.org/10.3390/app13020990
106. Han Z, Tian C, Zhou Z, Yuan Q (2022) Discovery of key function module in complex mechanical 3D CAD assembly model for design reuse. Assem Autom 42:54–66. https://doi.org/10.1108/AA-06-2021-0073
107. Stavropoulos P, Tzimanis K, Souflas T, Bikas H (2022) Knowledge-based manufacturability assessment for optimization of additive manufacturing processes based on automated feature recognition from CAD models. Int J Adv Manuf Technol 122:993–1007. https://doi.org/10.1007/s00170-022-09948-w

108. Yuan Y, Huh JH (2018) Customized CAD modeling and design of production process for one-person one-clothing mass production system. Electron 7.https://doi.org/10.3390/electronics7110270

109. Manavis A, Minaoglou P, Efkolidis N, Kyratsis P (2024) Digital customization for product design and manufacturing: a case study within the furniture industry. Electron. https://doi.org/10.3390/electronics13132483

110. Binder JR, Unver E, Benincasa-Sharman C et al (2024) Investigation of a new framework for mass customization within healthcare orientated human head data collection for healthcare professionals. Comput Aided Des Appl 21:499–509. https://doi.org/10.14733/cadaps.2024.499-509

111. Gembarski PC, Gast P (2024) Functional requirements and design features for the implementation of 3D CAD-based graphical interactive configurators. Appl Sci 14.https://doi.org/10.3390/app14073113

Chapter 2
Automated Design of Mechanical Parts: A Case Study

Abstract Design based on Computer-Aided Design (CAD) programming is a technique that relies on the programming of typical CAD systems via an integrated interface. The purpose of this technique is to automate certain engineering tasks such as the part design, the assembly, the drawing preparation and the G-code generation. The present paper focuses on the development of a code that is able to automate the design process of a simple mechanical part, by programming with the Application Programming Interface (API) of SolidWorks™ CAD software. The paper presents the developed User Interface (UI) and the written code for the case study. In addition, it illustrates the process and visualizes the given results. Visual Basic for Applications (VBA™) programming language was implemented to develop the code, mostly because it requires objects and enables a graphical interaction with them. The final product represents a typical steel angle used in the industry, which can be designed in a variety of profile dimensions and lengths.

Keywords Application programming interface (API) · Automated design · Computer-aided design (CAD) · CAD-based design · SolidWorks™ · Visual basic for applications (VBA)

2.1 Introduction

Computer-Aided Design (CAD) has revolutionized the fields of engineering, architecture, and product design by providing powerful tools for the creation, modification, analysis, and optimization of designs. CAD-based programming, an extension of traditional CAD, leverages programming languages and computational methods to automate, enhance, and innovate design processes.

CAD systems have evolved significantly since their inception in the 1960s. Initially, these systems were primarily used for drafting and documentation. However, with advancements in computing power and software capabilities, CAD has transformed into an essential tool for 3D modeling, simulation, and visualization [1]. The integration of programming into CAD systems marks a significant milestone, enabling the creation of parametric and algorithm-driven designs that are both precise and highly customizable [2]. CAD-based programming involves the use of scripting languages, such as Python, JavaScript, and domain-specific languages, to control and automate CAD operations. These languages facilitate the creation of custom functions and routines that can manipulate CAD models, generate complex geometries, and perform batch processing tasks [3].

Parametric design, a key concept in CAD-based programming, allows designers to define relationships and constraints among different elements of a model. Changes to one parameter automatically propagate through the model, ensuring consistency and reducing manual rework [4, 5]. Algorithmic design involves the use of algorithms to generate and manipulate geometric forms. This approach is particularly useful in creating complex, nonrepetitive patterns and structures that would be difficult to achieve manually [6]. Generative design uses algorithms to explore a vast design space and generate optimized solutions based on predefined criteria and constraints. This method leverages the computational power to identify innovative and efficient designs [7].

CAD-based programming is applied across various domains, offering significant advantages in terms of efficiency, accuracy, and innovation. In architecture, CAD-based programming is used to create parametric models that can adapt to different design requirements and site conditions. This approach allows designers and engineers to explore a wide range of design alternatives and optimize building performance. Engineering disciplines, including mechanical and civil engineering, benefit from CAD-based programming through the automation of repetitive tasks, such as the generation of standard components and the execution of finite element analyses. In product design, CAD-based programming enables the rapid prototyping of complex parts, facilitating iterative testing and refinement. This capability is particularly valuable in industries such as automotive and aerospace, where precision and performance are critical. The construction industry leverages CAD-based programming for the design and fabrication of custom building elements, such as facade panels and structural components. This technology supports the efficient production of bespoke solutions that meet specific project requirements.

Especially for product design and manufacturing, CAD-based programming is essential, since automation is the primary concern in such studies. Tzotzis et al. [8] developed a CAD-based application for the automatic generation of CNC programs [9], by utilizing solid modeling feature recognition techniques. Studies that utilize similar techniques deal with the design of various products and systems, such as furniture, bicycle and boat hull [10–12]. In addition, the utilization of CAD API in manufacturing is also evident. Wang and Chen [13] managed to predict the cutting tool defection accurately with the developed algorithm, based on the CAD integration with Computer-Aided Manufacturing (CAM) and Computer-Aided Engineering (CAE) as well. The advantages that derive from the interconnection between CAx systems were employed in studies related to the simulation of machining processes, the measurement and verification of machined surfaces, as well as in the topography evaluation of face-milled surfaces [14–16].

In the present study, an effort was made to highlight the significance of the CAD-based programming by presenting the development of a design automation tool. The VBA™ programming language was used to write the necessary code, while the typical API methods for the sketch creation, the extrude, as well as the revolve commands were implemented in the code. Finally, a comparison was made between a design table and the developed tool in terms of usability, processing speed, and editing convenience, in order to evaluate the advantages of the developed tool.

2.2 Material and Methods

2.2.1 User Interface

To facilitate the interaction between the user and the software [17], the User Interface (UI) shown in Fig. 2.1 was developed. Usually, a UI includes several textboxes for the variable input, command buttons for the execution of a certain procedure and sometimes option buttons for selecting different options. The specific UI contains three textboxes: one for the profile dimensions input, one for the length input and one for the revolve angle input. Moreover, a command button for the execution of the process is available, as well as another one for terminating the application and exiting the UI. Finally, two option buttons are present, so that the user can select between the straight and the curved version of the steel angle. A preview image of the product adds to the visualization and informs the user about the dimensions. In addition, the labels are used to describe the purpose of a graphical object, as well as to add useful information.

2.2.2 Design Process Flow

The design process flow for creating the 3D part in SolidWorks™ CAD software starts by defining certain variables and proceeds through a series of steps to create and save the solid part.

First, declaration of variables (i.e., dimensions) corresponds to the necessary inputs, such as Extrude length, Revolve angle, Profile length, Thickness, and Edge radius. These are the key dimensions or parameters that will be used throughout the design process. They define the basic attributes of the 3D model to be created. Next, the design options are matched with the set variables. Based on the declared variables, design options that meet these criteria are identified. This ensures that the subsequent steps will produce a model that adheres to the specified dimensions and requirements. The creation of new part document means that a new document is created within the CAD software to house the design. This document will contain all the elements and steps of the 3D part creation process. The fourth step is the plane selection. A plane is

Fig. 2.1 UI of the steel
angle design application

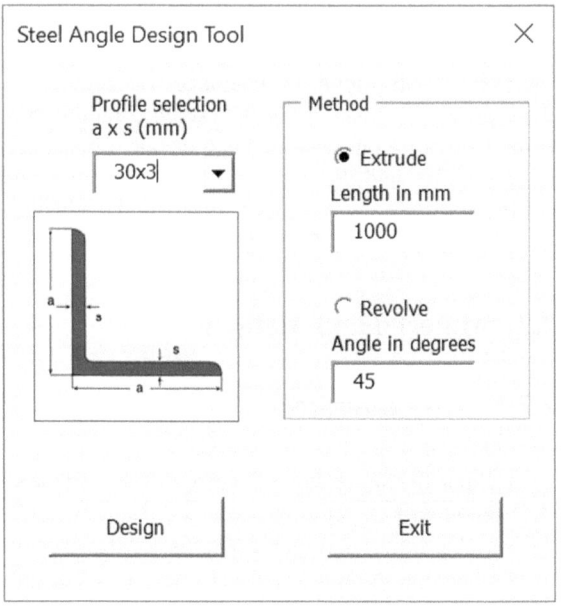

selected within the CAD environment on which the initial sketch will be drawn. This
plane serves as the foundational reference for the entire design. Subsequently, a new
sketch is created. Specifically, a new 2D sketch is created on the selected plane. This
sketch will serve as the base outline for the 3D model. The sixth step is the profile
creation. The sketch is refined and detailed to create a specific profile. This profile
represents the cross-sectional shape of the part (steel angle) that will be extruded or
revolved. The seventh step is the application of the appropriate feature (Extrude or
Revolve). Depending on the design requirements, either an extrusion or a revolve is
applied to the profile. The extrude extends the profile linearly to create a 3D shape,
therefore a straight steel angle in this case. On the contrary, revolve rotates the profile
around an axis to create a 3D shape. The final 3D solid part is saved during the final
step. This completes the design process and ensures that the part can be accessed,
modified, or manufactured later.

Figure 2.2 illustrates the process flow in a manner typical for the automated design
with SolidWorks™ [18]. The process starts with declaring the necessary variables,
which set the foundation for the design. It then moves to identifying design options
that conform to these variables. To start the design, first a new part document is
created, and then a plane is selected. A sketch is created on the selected plane, refined
into a profile, and then transformed into a 3D model through extrusion or revolution.
Finally, the solid part is saved, completing the design process. This structured flow
ensures that the design process is methodical and adheres to predefined dimensions,
resulting in a precise and accurate 3D model.

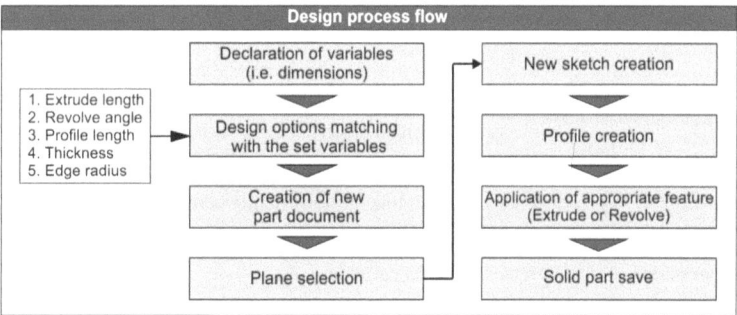

Fig. 2.2 Process flow during the part design

2.2.3 Process Code

The full code was written with VBA™ and is available in the Appendix section. To use the code, it must be put inside a SolidWorks™ macro file, the form shown in Fig. 2.1 must be created and finally, the appropriate matching between the form elements names (i.e., command buttons, text boxes) in the code and in the properties window must be done.

By observing the code, a reasonable flow is evident. First, the variables are declared and are linked to the corresponding text boxes. Next, the connection with SolidWorks™ is established and a new part document is created with the "New-Document" method. Subsequently, the default plane is selected with the "Select-ByID2" method, which can be used to select many geometrical objects such as planes, surfaces, points, and lines [19]. Then, a new sketch is inserted and the part profile is designed. The "InsertSketch" method is utilized to open a new sketch on the selected plane, whereas the "CreateLine" is used to design a set of lines. In this case, the lines were used to form the steel angle profile. Finally, the appropriate feature is applied to the sketch, according to whether the Extrude or Revolve option is enabled, generating the final solid model of the product. The "if-then-else" statement is used to enable the feature selection. To create the straight solid model, "Feature-Extrusion3" method is used. On the other hand, "FeatureRevolve2" is responsible for the revolved feature. This feature requires an axis around which, the solid model is generated. Therefore, the "CreateCenterLine" method is used to create such an axis. Table 2.1 summarizes the typical API methods used to synthesize the developed tool's code.

Table 2.1 Basic API methods for the solid part creation [20]

Method	Explanation
NewDocument	Creates a new document based on the specified template
SelectByID2	Selects the specified entity
InsertSketch	Inserts a new sketch in the current part or assembly document
CreateLine	Creates a sketch line in the currently active 2D or 3D sketch
CreateCenterLine	Creates a center line between the specified points
FeatureExtrusion3	Creates an extruded feature
FeatureRevolve2	Creates a base-, boss-, or cut-revolve feature

2.3   Results

The final result after the design process is finished, is a complete solid model of either a straight steel angle or a revolved steel angle with the preferred dimensions. Figure 2.3 depicts the two models.

To evaluate the performance of the developed tool in terms of the time and effort required to write the code (Initial setup), to select the dimensions and available options (Variable input), to perform the design process (Processing), and to modify the code so that the tool can handle similar parts (Extending), Table 2.2 was prepared. In addition, a comparison between the tool's and the design table's performances was made. For the design table, the same performance parameters apply. Therefore, "Initial setup" corresponds to the preparation of the table. "Variable input" refers to the variable allocation and the matching with each dimension. "Processing" is the selection and build of the already designed configurations and finally "Extending" would correspond to the process with which the table could be enriched with extra features. This however, is not possible. The indexes 0 to 5 used in the table correspond to the next statements: 0 = No effort (work/processing time is virtually nonexistent); 1 = Very little effort (work/processing time under 5 min); 2 = Little effort (work/processing time between 5 and 15 min); 3 = Normal effort (work/processing time between 15 to 30 min); 4 = Strong effort (work/processing

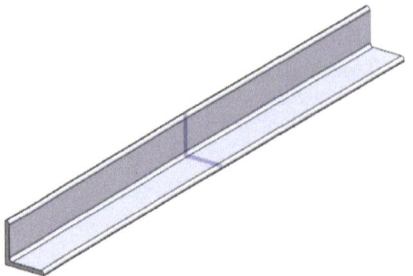

Fig. 2.3 Two types of steel angles available

Table 2.2 Evaluation parameters and their indexes for the performance comparison (the lower, the better)

Parameter	Index	
	Developed tool	Design table
Initial setup	4	3
Processing	0	1
Material editing	1	2
Extending	2	n/a

time approximately 60 min); 5 = Very strong effort (work/processing time more than 60 min). All the indexes used are indicative. However, they were measured by utilizing the API methods and the design table tools of SolidWorks™ CAD system by experienced users.

It is evident that the tool requires more time to be prepared and set, compared to a design table. Nevertheless, once set, it provides better performance both in processing and material editing. Moreover, it allows for the extension and upgrading of the tool's capabilities, whereas the design table does not have such option at all.

2.4 Conclusions

The paper showed that utilizing CAD API enhances productivity and optimizes work time allocation. This can be achieved by developing tools or applications, with the API, tailored to specific activities, eliminating the need to spend extensive time coding or learning more high-level programming languages. When these tools or applications can automate a variety of desired tasks with minimal input parameters, they become invaluable assets in the daily work of designers and engineers alike.

The present study is based on the CAD API to develop a tool for automating the design process of standardized steel angles, demonstrating the use of the extrude and revolve methods. Moreover, by analyzing the coding process, the next remarks can be concluded:

- The implementation of CAD programming in the design process can help automate routine engineering tasks involved in such processes.
- It is possible to extend the usability of the tool by editing the code, enabling the design of similar standardized parts, such as other types of steels, with minimal effort.
- Finally, the development of automation tools and applications can eliminate the need for design tables and other traditional CAD tools that require more effort and time to set, edit, or modify.

Appendix

```vb
'******************************************************
' Steel angle design tool
'******************************************************
 Option Explicit

' Declaration of SolidWorks application object and docu-
ment object
 Public swApp As SldWorks.SldWorks
 Public swPart As SldWorks.PartDoc

' Variable declaration
 Dim boolstatus As Boolean
 Dim swSketch As Object
 Dim skSegment As Object
 Dim Length As Double
 Dim Angle As Double
 Dim a, s As Double
 Dim swFeature As SldWorks.Feature
 Dim pi As Double


'------------------------------------------------------
' Start of code for Build command button
'------------------------------------------------------
Private Sub Build_Button_Click()

' 1. Variable values definition
  pi = 4 * Atn(1)
  Length = CDbl(Length_TextBox.Text) / 1000
  Angle = CDbl(Angle_TextBox.Text * pi / 180)

' 2. Connection to SolidWorks
 Set swApp = Application.SldWorks
 Set swPart = swApp.ActiveDoc

' 3a. Disable the ability to add manually a dimension
value
 swApp.SetUserPreferenceToggle swInputDimValOnCreate,
False

' 3b. New Part document creation
 Set swPart = swApp.NewDocument _
 ("C:\ProgramData\SolidWorks\SolidWorks
2023\templates\Part.prtdot", 0, 0, 0)
' 4a. Sketch creation on Front plane
   boolstatus = swPart.Extension.SelectByID2("Front
 Plane", "PLANE", 0, 0, 0, False, 0, Nothing, 0)
```

```vba
  swPart.SketchManager.InsertSketch True
  Set swSketch = swPart.SketchManager.CreateLine(0, 0, 0,
0, a, 0)
  swPart.AddDimension2 -s, a / 2, 0
  Set swSketch = swPart.SketchManager.CreateLine(0, a, 0,
s, a, 0)
  swPart.AddDimension2 s / 2, a + s, 0
  Set swSketch = swPart.SketchManager.CreateLine(s, a, 0,
s, s, 0)
  Set swSketch = swPart.SketchManager.CreateLine(s, s, 0,
a, s, 0)
  Set swSketch = swPart.SketchManager.CreateLine(a, s, 0,
a, 0, 0)
  Set swSketch = swPart.SketchManager.CreateLine(a, 0, 0,
0, 0, 0)
' 4b. Creation of fillet
  boolstatus = swPart.Extension.SelectByID2("Line2",
"SKETCHSEGMENT", s / 2, a, 0, True, 0, Nothing, 0)
  boolstatus = swPart.Extension.SelectByID2("Line3",
"SKETCHSEGMENT", s, a / 2, 0, True, 0, Nothing, 0)
  Set skSegment = swPart.SketchManager.CreateFillet(s *
0.8, 1)
  boolstatus = swPart.Extension.SelectByID2("Line3",
"SKETCHSEGMENT", s, a / 2, 0, True, 0, Nothing, 0)
  boolstatus = swPart.Extension.SelectByID2("Line4",
"SKETCHSEGMENT", a / 2, s, 0, True, 0, Nothing, 0)
  Set skSegment = swPart.SketchManager.CreateFillet(s *
0.8, 1)
  boolstatus = swPart.Extension.SelectByID2("Line4",
"SKETCHSEGMENT", a / 2, s, 0, True, 0, Nothing, 0)
  boolstatus = swPart.Extension.SelectByID2("Line5",
"SKETCHSEGMENT", a, s / 2, 0, True, 0, Nothing, 0)
  Set skSegment = swPart.SketchManager.CreateFillet(s *
0.8, 1)
' 5. Extrude or Revolve feature application
 If Extrude_Button = True Then
  Set swFeature =
swPart.FeatureManager.FeatureExtrusion3(True, False,
False, 6, 0, Length, 0, False, _
  False, False, False, 0, 0, False, False, False, False,
True, True, True, 0, 0, False)
 Else
  Set swSketch = swPart.SketchManager.CreateCenterLine(-2
* a, 0, 0, -2 * a, a, 0)
  boolstatus = swPart.Extension.SelectByID2("Line7",
"SKETCHSEGMENT", 0, 0, 0, True, 0, Nothing, 0)
```

```
   Set swFeature =
swPart.FeatureManager.FeatureRevolve2(True, True, False,
False, False, False, 0, 0, _
  Angle, 0, False, False, 0, 0, 0, 0, 0, True, True,
True)
 End If
' 6. Model save
 Set swPart = swApp.ActiveDoc
 swPart.Extension.RunCommand swCommands_SaveAs, Empty
End Sub
'-----------------------------------------------------------
' End of code for Build command button
'-----------------------------------------------------------

' Exit form
Private Sub Exit_Button_Click()
 End
End Sub
' Form initialization and default values
Private Sub UserForm_Initialize()
 Extrude_Button = True
 Length_TextBox.Text = 1000
 Angle_TextBox.Text = 45
 Dim_ComboBox.Text = "30x3"
' Combo Box filling
 With Dim_ComboBox
 .AddItem "30x3"
 .AddItem "40x4"
 .AddItem "50x4"
 .AddItem "50x5"
 .AddItem "60x5"
 .AddItem "60x6"
 .AddItem "70x7"
 .AddItem "70x8"
 End With
End Sub
' Steel angle standardized options
Private Sub Dim_ComboBox_Change()
 If (Dim_ComboBox.Value = "30x3") Then
  a = 30 / 1000: s = 3 / 1000
 ElseIf (Dim_ComboBox.Value = "40x4") Then
  a = 40 / 1000: s = 4 / 1000
 ElseIf (Dim_ComboBox.Value = "50x4") Then
   a = 50 / 1000: s = 4 / 1000
 ElseIf (Dim_ComboBox.Value = "50x5") Then
   a = 50 / 1000: s = 5 / 1000
```

```
  ElseIf (Dim_ComboBox.Value = "60x5") Then
     a = 60 / 1000: s = 5 / 1000
  ElseIf (Dim_ComboBox.Value = "60x6") Then
     a = 60 / 1000: s = 6 / 1000
  ElseIf (Dim_ComboBox.Value = "70x7") Then
     a = 70 / 1000: s = 7 / 1000
  ElseIf (Dim_ComboBox.Value = "70x8") Then
     a = 70 / 1000: s = 8 / 1000
  End If
End Sub
'***********************************************************
' End of program
'***********************************************************
```

References

1. Antoine Brière-Côté LR, Maranzana R (2012) Comparing 3D CAD models: uses, methods, tools and perspectives. Comput Aided Des Appl 9:771–794. https://doi.org/10.3722/cadaps. 2012.771-794
2. Qin Y, Wang C (2024) Intelligent algorithm-driven product design process optimization: intelligent transformation of product design processes. Appl Math Nonlinear Sci 9:1–17
3. Frank G, Entner D, Prante T, Schwarz M (2014) Towards a generic framework of engineering design automation for creating complex CAD models. Int J Adv Syst Meas 7:179–192
4. Camba JD, Contero M, Company P (2016) Parametric CAD modeling: an analysis of strategies for design reusability. Comput Des 74:18–31. https://doi.org/10.1016/j.cad.2016.01.003
5. Gu N, Yu R, Behbahani PA (2021) Parametric design: theoretical development and algorithmic foundation for design generation in architecture. In: Sriraman B (ed) Handbook of the mathematics of the arts and sciences. Springer International Publishing, Cham, pp 1361–1383
6. Castelo-Branco R, Caetano I, Leitão A (2022) Digital representation methods: the case of algorithmic design. Front Archit Res 11:527–541. https://doi.org/10.1016/j.foar.2021.12.008
7. Ntintakis I, Stavroulakis GE (2020) Progress and recent trends in generative design. MATEC Web Conf 318:01006. https://doi.org/10.1051/matecconf/202031801006
8. Tzotzis A, Manavis A, Efkolidis N, Kyratsis P (2021) CAD-based automated G-code generation for drilling operations. Int J Mod Manuf Technol 13:177–184. https://doi.org/10.54684/ijmmt. 2021.13.3.177
9. Kumar K, Ranjan C, Davim JP (2020) CNC programming for machining. Springer Nature Switzerland
10. Kyratsis P, Tzotzis A, Manavis A (2021) Computational design and digital fabrication. In: Kumar S, Rajurkar KP (eds) Advances in manufacturing systems. Springer Singapore, Singapore, pp 1–16
11. Manavis A, Tzotzis A, Tsagaris A, Kyratsis P (2022) A novel computational-based visual brand identity (CbVBI). Machines
12. Tzotzis A, Efkolidis N, García-Hernández C, Kyratsis P (2024) CAD-based automated USV hull design. In: Guxho G, Kosova Spahiu T, Prifti V et al (eds) Proceedings of the joint international conference: 10th textile conference and 4th conference on engineering and entrepreneurship. Springer Nature Switzerland, Cham, pp 354–362
13. Wang L, Chen ZC (2014) A new CAD/CAM/CAE integration approach to predicting tool deflection of end mills. Int J Adv Manuf Technol 72:1677–1686. https://doi.org/10.1007/s00 170-014-5760-4
14. Tzotzis A, Tsagaris A, Tapoglou N, Kyratsis P (2023) High-precision CAD-based simulation for turning considering tool microgeometry. Int J Mechatron Manuf Syst 16:83–95. https://doi. org/10.1504/IJMMS.2023.132023

15. Gella-Marín R, Tzotzis A, García-Hernández C, et al (2021) CAD software integration with programming tools for modelling, measurement and verification of surfaces. In: Experiments and simulations in advanced manufacturing, pp 91–116
16. Tapoglou N, Efstathiou C, Tzotzis A, Kyratsis P (2023) Study of the topography of face milled surfaces using CAD-based simulation. In: Kyratsis P, Manavis A, Davim JP (eds) Computational design and digital manufacturing. Springer International Publishing, Cham, pp 159–166
17. Tzotzis A, García-Hernández C, Huertas-Talón JL, Kyratsis P (2020) CAD-based automated design of FEA-ready cutting tools. J Manuf Mater Process 4:1–14. https://doi.org/10.3390/jmmp4040104
18. Tzotzis A, Garcia-Hernandez C, Huertas-Talon J-L et al (2017) Engineering applications using CAD based application programming interface. In: MATEC web of conferences, pp 1–7
19. Kyratsis P, Gabis E, Tzotzis A et al (2019) CAD based product design: a case study. Int J Mod Manuf Technol 11:88–93
20. Kyratsis P, Tzotzis A, Tzetzis D, Sapidis N (2018) Pneumatic cylinder design using cad-based programming. Acad J Manuf Eng 16:107–113

Chapter 3
Solid Model Geometric Objects Detection Based on CAD Software Programming

Abstract Computer-Aided Design (CAD) programming can be integrated into many engineering procedures that ranges from design and assembly of mechanical parts and products, to manufacturing. This work demonstrates the advantages of the Application Programming Interface (API) of SolidWorks™ CAD software in terms of topology detection that can be utilized in many routine tasks during the development phase of a product. These characteristic processes are embedded into a tool that is able to automatically apply typical design features such as chamfer, fillet, and hole to the corresponding geometric object (edge or face). The tool operates with minimal input from the user, requiring only the name of the geometric object in the case of either the chamfer or fillet and the center coordinates, in addition to the name of the appropriate geometric object, in the case of the hole feature. It is noted that the code was written with the Visual Basic for Applications (VBA™) language and can be applied to any solid model.

Keywords Application programming interface (API) · Computer-aided design (CAD) · CAD-based programming · Geometric features · Solid models · SolidWorks™ · Geometric objects · Visual basic for applications (VBA)

3.1   Introduction

Solid models are a crucial element in the design and manufacturing process. They are used to represent objects in three dimensions, providing detailed information about their shape, structure, and function. Computer-Aided Design (CAD) software is pivotal in creating these models, allowing engineers and designers to visualize and manipulate the physical properties of objects before they are produced. An essential aspect of working with solid models is the ability to recognize geometric features, a process that can be automated using programming techniques within CAD software. Geometric Feature Recognition (GFR) is vital for various applications, including manufacturing, finite element analysis, and reverse engineering.

© The Author(s), under exclusive license to Springer Nature Switzerland AG 2025   39
P. Kyratsis et al., *CAD-based Programming for Design and Manufacturing*,
Manufacturing and Surface Engineering, https://doi.org/10.1007/978-3-031-78747-8_3

Geometric features are specific shapes, such as holes, slots, pockets, or bosses, that are significant in defining the design and function of a part. Recognizing these features within solid models is essential for automating the design and manufacturing processes. GFR can assist in generating machining instructions, conducting stress analysis, and verifying design intent. Automating feature recognition can significantly reduce the time and cost associated with product development. The integration of GFR in CAD systems can lead to more efficient and accurate design processes. By automating the recognition of features, designers can focus on more complex aspects of the design, leaving repetitive tasks to the software. Furthermore, GFR can facilitate the transition from design to manufacturing by automatically identifying features that need to be machined, thus streamlining the process of generating tool paths. There are various approaches to implementing geometric feature recognition in CAD systems. These methods can generally be classified into three categories: rule-based, graph-based, and artificial intelligence-based approaches.

Rule-based methods [1] are among the earliest approaches to feature recognition. They rely on predefined rules and heuristics to identify features within a CAD model. For example, a rule might state that any cylindrical surface with a certain diameter and orientation is a hole. While this approach is straightforward and relatively easy to implement, it has limitations in dealing with complex features or models with a large variety of feature types. Additionally, rule-based methods often require extensive manual input to define the rules, making them less adaptable to new or modified designs. Studies have shown that rule-based methods can be effective for simple parts but struggle with more complex geometries.

Graph-based methods [2] represent the solid model as a graph, where nodes correspond to geometric entities (such as faces or edges), and edges represent the relationships between these entities. Feature recognition is achieved by searching for subgraphs within the model that match the topology of known features. This method is more flexible than rule-based approaches, as it can handle a wider variety of features without the need for extensive manual rule definition. Graph-based methods are particularly effective in recognizing features in complex models. According to a study by Venkataraman et al. [3], graph-based techniques offer a more robust solution for feature recognition in CAD systems, especially when dealing with intricate geometries and multiple feature types. The ability to decompose a model into subgraphs allows for a more modular and scalable approach to feature recognition.

The advent of Artificial Intelligence (AI) and machine learning has opened new avenues for geometric feature recognition. AI-based methods use training data to learn the characteristics of different features, allowing them to recognize features in new models without explicit programming. Machine learning algorithms, such as neural networks and decision trees, have been applied to GFR with promising results. One of the main advantages of AI-based methods is their ability to adapt to new and complex features [4, 5]. As CAD models become more sophisticated, the flexibility of AI becomes increasingly important. An early research by Han and Requicha [6] demonstrated the potential of AI in feature recognition, particularly in its ability to learn from examples and improve accuracy over time.

Implementing geometric feature recognition in CAD software requires a combination of geometric reasoning, data structures, and algorithmic design. Modern CAD systems, such as SolidWorks™, AutoCAD™, and CATIA™, provide Application Programming Interfaces (APIs) that allow developers to create custom feature recognition tools. APIs in CAD software enable programmers to access the underlying geometric data of a model and apply algorithms for feature recognition. For example, the SolidWorks™ API provides functions to query geometric properties of solid bodies, such as volume, surface area, and feature types. By leveraging these functions, developers can create custom scripts or plugins that automate the recognition of features like holes, fillets, and chamfers. One approach to feature recognition using APIs is to iterate through the faces and edges of a solid model, checking for conditions that match known features [7]. For example, a cylindrical face with two circular edges might be recognized as a hole [8]. The use of APIs allows for the creation of sophisticated feature recognition tools that can be tailored to specific industries or applications. Research by Sunil and Pande [9] highlights the effectiveness of using AI for feature recognition, particularly in the context of manufacturing. The study demonstrates how custom tools can be developed to identify machining features [10] directly from CAD models, significantly reducing the time required to prepare models for production. Studies have suggested that a multifaceted approach is often necessary for effective feature recognition. For instance, Guo et al. [11] proposed a hybrid method that integrates graph-based techniques with rule-based heuristics, resulting in improved accuracy and performance in recognizing complex features. Such methods have found application in studies related to robotic planning, computational design, and manufacturing simulations [12–15] as well.

In the present study, the opposite process to the GFR is performed. Thus, instead of recognizing the design features that are already present on a model, the principal geometric objects are identified. In specific, the API-based geometric object handling method is used to identify two types of the primary geometric objects on solid parts. Specifically, edges and faces. The purpose is the automated creation of design features such as chamfer, fillet, and hole. Therefore, an appropriate tool was developed to house the specific processes and to demonstrate the usability of the automated geometric object selection in CAD systems.

3.2 Material and Methods

3.2.1 User Interface

Figure 3.1 illustrates the User Interface (UI) designed for creating and modifying geometric features, such as chamfers, fillets, and holes, on 3D models within a CAD environment. The UI is organized into two main sections, each allowing detailed customization of the features being added to the model. The first one allows the user to interact with certain inputs for the automated creation of features such as chamfer and fillet on edges and corners respectively. It contains a numerical input field where the user specifies the distance for a chamfer or the radius for a fillet. For example, entering "2" sets the distance from the edge for a chamfer or the radius of the curved surface for a fillet. Another field allows the user to input the angle for the chamfer. The angle defines the slant of the chamfered surface relative to the edge. For instance, entering "45" creates a 45-degree chamfer. A text field where the user inputs the name of the edge on which the feature will be applied. This could be any edge identified in the CAD model, such as "E1". The edge name ensures that the feature is applied to the correct part of the model. It should be noted that the naming of the corresponding geometric objects is performed automatically once the user activates the process. Two radio buttons enable the user to choose between creating a chamfer or a fillet. Selecting "Chamfer" configures the tool to apply the parameters as a chamfer, while "Fillet" sets the tool to create a rounded edge instead. Once all parameters are set, the user clicks the command button to apply the chamfer or fillet to the specified edge. The feature is then added to the 3D model according to the inputs provided.

The second section relates to the hole creation feature. In a similar manner it contains several input elements. The text field is where the user specifies the name of the face where the hole will be created. For example, entering "F1" identifies the specific face in the CAD model that will be drilled. A numerical input where the user defines the diameter of the hole to be cut. This determines the size of the circular opening in the face. For instance, entering "10" creates a hole with a 10 mm diameter. The depth input field specifies the depth of the hole to be drilled into the face. This defines how deep into the material the hole will penetrate. For example, entering "25" indicates a 25 mm deep hole. Finally, a numerical input field serves for precise positioning of the hole's center. The x, y, and z fields allow the user to input the exact coordinates where the hole should be placed. After setting the hole's dimensions and position, the "Cut Hole" button is used to execute the drilling operation. The hole is then created on the specified face with the defined parameters.

Overall, this UI provides a user-friendly interface [16] for adding chamfers, fillets, and holes to a 3D model. It offers precise control over each feature's dimensions and positioning, enabling users to customize their designs with accuracy and ease. The separation into distinct sections ensures clarity and focus when working with different types of features, making the process of model modification intuitive and efficient.

3.2.2   Geometric Object Detection, Traversing, and Selection

To describe the code's flow for the geometric object detection, traversing, and selection, the diagram of Fig. 3.2 was created. This diagram outlines the automated design feature creation, which is a process in a CAD software that automates the creation of specific features (chamfers, fillets and holes) on 3D models. The process starts with accessing the model and proceeds through steps that enables the application of specific features to selected parts of the model in a similar manner found in the

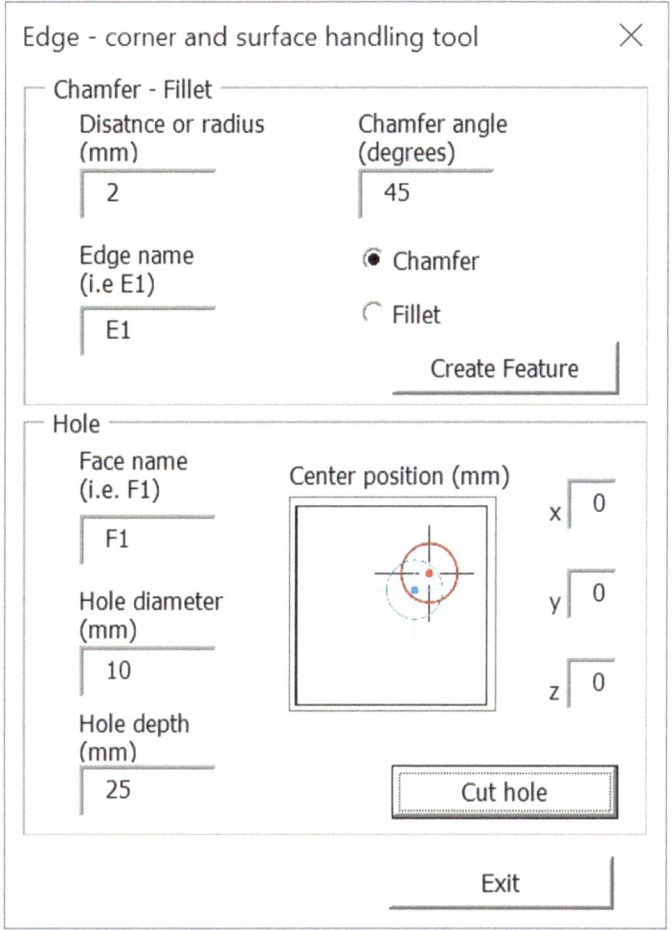

Fig. 3.1 UI of the geometric objects handling tool

study by Tzotzis et al. [17]. First step involves activating the current model document within the CAD software. It ensures that the correct 3D model is in focus and ready for modifications or feature additions. Next, the software retrieves the model document, which contains all the necessary data about the 3D model. This step ensures that the system has the necessary information to proceed with feature creation. The system accesses the solid bodies within the model. Solid bodies are the 3D components or parts of the model that can be modified or to which features can be added. The system identifies and retrieves the faces of the solid bodies. Faces are the flat or curved surfaces on the solid body that can be selected for applying features such as holes and fillets. The system retrieves the edges of the solid bodies. Edges are the lines where two faces meet and can be selected for operations like creating fillets or chamfers. It is possible to make some minor changes to the code so that it can identify and retrieve both edges and faces at the same time. Subsequently, the geometry objects (faces or edges) that were retrieved in the previous step are named or labeled within the system. This naming helps in identifying and selecting the appropriate objects for the next steps. The system selects specific objects (faces or edges) from the named objects, according to the one given by the user in the appropriate textbox. This selection determines where the feature will be applied. If the selected objects are appropriate for creating a hole, the hole feature is applied, which essentially is a cut extrude feature. The software automatically generates a hole in the selected face according to predefined parameters. Else, if the selected objects are edges or are suitable for a fillet or chamfer, this feature is applied. The software automatically rounds off (fillet) or bevels (chamfer) the selected edge.

Summarizing, the process begins by ensuring that the correct 3D model is active and accessible. The system then retrieves relevant parts of the model, such as faces or edges, that could be modified. After retrieving and naming these objects, the system selects specific areas where features like holes, fillets, or chamfers should be applied. Finally, the selected features are automatically applied to the model, streamlining the design process. This flow is particularly useful for automating repetitive tasks in CAD software, ensuring consistency and efficiency in applying standard features across a 3D model.

3.2.3 Process Code

The complete code is available in the Appendix section, which requires to be put in a macro file, in order to become executable [18, 19]. Moreover, the form shown in Fig. 3.1 is also necessary, along with the appropriate variable matching, to allow the user interaction. Table 3.1 includes the most important API methods for the automated design feature creation.

First "ActiveDoc" method is required for the connection between the part document and the API hierarchy. The "GetBodies2", "GetFaces", and "GetEdges" methods are responsible for the solid body access, the retrieval of the faces, and the edges respectively. The faces and edges that are available in the solid bodies are traversed with a loop and their data are stored to an array. Moreover, each geometric object is named with the "SetEntityName" method by using a specified prefix (i.e., E, from edge) and a numerical suffix that follows the order of a counter (i.e., 1, 2, 3, and so on). Methods "GetEntityByName" and "Select4" are used to select the specified object. The selection does not differ from the normal selection with the mouse pointer in the CAD environment. Finally, the methods "InsertFeatureChamfer" and "FeatureFillet3" are required for the chamfer and fillet features respectively. To create a hole on the selected geometry, the "FeatureCut4" method is utilized.

The variables that are required for the solid body and geometric object traverse loop, as well as for the registration of the found geometry objects, are shown in Table 3.2. As already mentioned, the process relies on the detection of the solid

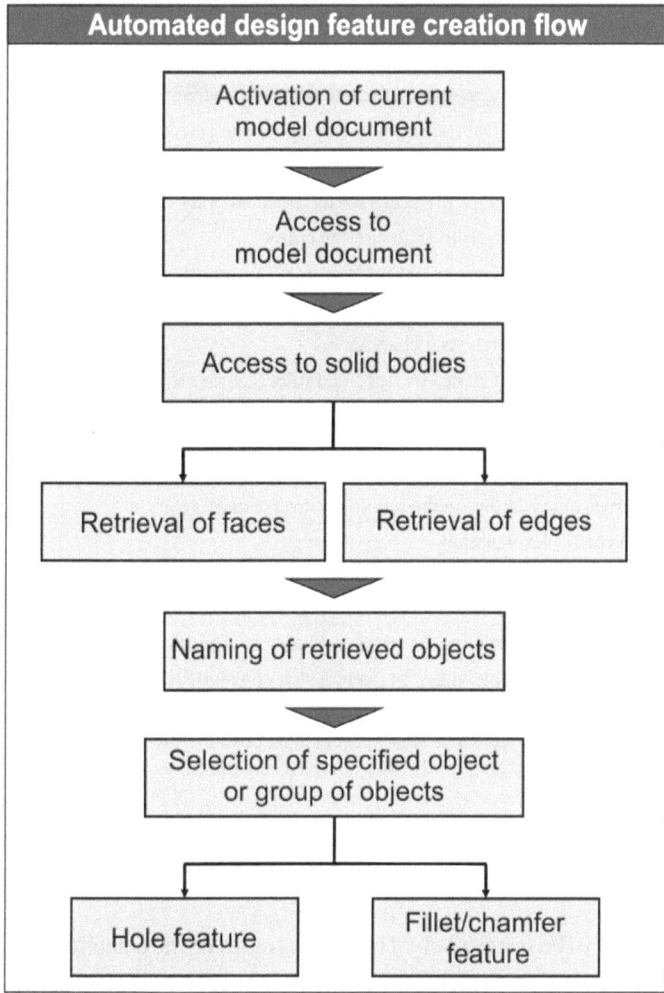

Fig. 3.2 Process flow during the geometric object selection and handling

bodies that are available in the part document, the geometric objects that are available on the solid bodies and the registration of each of the found objects for later use.

3.3 Results

A simple orthogonal solid part was used to test the developed tool. All three design features (chamfer, fillet and hole) were tested successfully, as well as the naming procedure. All twelve edges and six faces were named accordingly, beginning with

Table 3.1 Basic API methods for the geometric object selection and design feature creation

Method	Explanation
ActiveDoc	Connects to the currently active document
GetBodies2	Gain access to the bodies in the currently active part
GetFaces	Gets all the faces on the body
GetEdges	Gets the edges for the selected body
SetEntityName	Sets the name of the entity
GetEntityByName	Gets an entity (face, edge, vertex) by name
Select4	Selects an entity and marks it
InsertFeatureChamfer	Inserts a chamfer
FeatureFillet3	Creates the specified fillet feature for selected edges or faces
FeatureCut4	Creates a cut extrude feature

Table 3.2 Variables required during the topology traverse and detection

Variable	Type	Explanation
vBodies	Variant	Variable for the solid body, available in the part document, data assignment
vEdges	Variant	Variable for the edges, available in the solid body, data assignment
i	Integer	Variable to assign integers that constitute the beginning and end of the loop used to detect the solids that are present in the document
j	Integer	Variable to assign integers that constitute the beginning and end of execution of the loop for detecting the edges that are present in the solid body
EdgeCounter	Integer	Variable to assign integers that make up the edge count, which is essentially the execution number of the edge detection loop

E1 to E12 and F1 to F6 respectively. The process relies on the detection and traversing of the target geometric objects group, as well as the naming of each one of the objects discovered. Thus, the selection of the preferred object and the application of the required feature becomes simplified. It is noted that the geometric object naming is not a dynamic procedure, meaning that every time a new face or edge is created due to a chamfer, fillet or hole, it will not be given a name once the code is executed anew. This way, any errors in naming such as double name, can be avoided. Moreover, the naming is persistent once the model is saved, meaning that the names of the edges and faces will not be erased, even after the session is closed. Finally, the order in which the names were given, remains the same no matter the number of the code's execution. Next, the most important advantages that derive from this method, after testing a simple part, are summarized:

- Error-free, static naming process.
- Persistent geometric object names.
- Specific naming order that does not change.

A final remark is the fact that it is possible to add another command button, which can be used to execute the naming process, prior to the application of any kind of design feature. This alteration requires minor editing to the code; however, it does not add any particular benefit, rather it might be more practical when the user wishes to only name the geometric features.

3.4 Conclusions

GFR is a critical component of modern CAD systems, enabling the automation of many tasks that are essential to the design and manufacturing process. While traditional methods, such as rule-based and graph-based approaches, have been effective, the advent of AI and machine learning offers new possibilities for more flexible and adaptive feature recognition. The integration of GFR with other technologies, such as augmented reality and virtual reality, could provide new ways for designers and engineers to interact with and visualize features within solid models. These advancements have the potential to revolutionize the way geometric feature recognition is performed, leading to more efficient and innovative design processes. By leveraging CAD software programming and APIs, developers can create custom tools that improve the efficiency and accuracy of design feature and geometric object recognition, ultimately leading to more effective product development.

Concluding, API-based programming remains a standard method to deal with automated processes within CAD environment. Moreover, it constitutes the basis for any other advancement in the field. The present study, demonstrated the use of the API-based method in geometric feature creation processes, enabling the quick and accurate detection of geometric objects within a solid part, leading to effortless application of design features.

Appendix

```vb
'***********************************************************
' Edge – corner and surface handling tool
'***********************************************************
Option Explicit
' Declaration of SolidWorks application object and docu-
ment object
Public swApp As SldWorks.SldWorks
Public swModel As SldWorks.ModelDoc2

' Variable declaration
Dim swPart As SldWorks.PartDoc
Dim vBodies As Variant
Dim swBody As SldWorks.Body2
Dim vEdges, vFaces As Variant
Dim swEdge As SldWorks.Edge
Dim swFace As SldWorks.Face2
Dim swEnt As SldWorks.Entity
Public swSelectData As SldWorks.SelectData
Dim swFeature As SldWorks.Feature
Dim swSketch As Object
Dim Diameter, Depth, Distance, Angle, pi, x, y, z As Dou-
ble

Dim i, j As Integer
Dim EdgeCounter, FaceCounter As Integer
'----------------------------------------------------------
' Start of code for Feature_Button_Click command button
'----------------------------------------------------------
Public Sub Feature_Button_Click()
' 1. Connection to SolidWorks
    Set swApp = Application.SldWorks
    Set swModel = swApp.ActiveDoc
    Set swPart = swModel

' 2. Variable values definition
  pi = 4 * Atn(1)
  Distance = CDbl(Distance_TextBox.Text) / 1000
  Angle = CDbl(Angle_TextBox.Text) * pi / 180

' 3. Check for open part document
 If swPart Is Nothing Then
 MsgBox ("Pleas open a part " & vbNewLine & _
 "document.")
```

```
    Exit Sub
  End If
' 4. "For" loop for naming edges
    EdgeCounter = 1

    vBodies = swPart.GetBodies2(swAllBodies, False)
    For i = 0 To UBound(vBodies)
        Set swBody = vBodies(i)
        vEdges = swBody.GetEdges
        For j = 0 To UBound(vEdges)
            Set swEdge = vEdges(j)
            Set swEnt = swEdge
            Debug.Print swPart.SetEntityName(swEnt, "E" &
EdgeCounter)
            EdgeCounter = EdgeCounter + 1
        Next j
    Next i

' 5. Selection of specified edge
        Set swEdge = swPart.GetEntityBy-
Name(UserForm1.TextBox1.Text, swSelEDGES)
        Set swEnt = swEdge
' 5b. Check for specified edge
      If swEnt Is Nothing Then
        MsgBox "Edge " & TextBox1.Text & " does not exist
or already have a chamfer/fillet." _
        & vbNewLine & "Please choose another."
        Exit Sub
      Else
        swEnt.Select4 True, swSelectData
      End If

' 6. Conditional structure for chamfer / fillet
    If Chamfer_Button = True Then
    Set swFeature = swPart.FeatureManager.InsertFea-
tureChamfer(2, 1, Distance, Angle, 0, 0, 0, 0)
    swPart.ClearSelection2 True
    Else
    Set swFeature = swPart.FeatureManager.Feature-
Fillet2(194, Distance, 0, 0, 0, 0, 0, 0, 0, 0)
    swPart.ClearSelection2 True
    End If

End Sub

'------------------------------------------------------
```

```vb
' End of code for Feature_Button_Click command button
'-------------------------------------------------------


'-------------------------------------------------------
' Start of code for Hole_Button_Click command button
'-------------------------------------------------------
Private Sub Hole_Button_Click()
' 1. Connection to SolidWorks
    Set swApp = Application.SldWorks
    Set swModel = swApp.ActiveDoc
    Set swPart = swModel

' 2. Check for open part document
 If swPart Is Nothing Then
  MsgBox ("Please open a part" & vbNewLine & _
  "document.")
  Exit Sub
 End If

' 3. Variables value definition
  Diameter = CDbl(Diameter_TextBox.Text) / 1000
  Depth = CDbl(Depth_TextBox.Text) / 1000
  x = CDbl(x_TextBox.Text) / 1000
  y = CDbl(y_TextBox.Text) / 1000
  z = CDbl(z_TextBox.Text) / 1000
' 4. "For" loop for naming faces
    FaceCounter = 1
    vBodies = swPart.GetBodies2(swAllBodies, False)
    For i = 0 To UBound(vBodies)
        Set swBody = vBodies(i)
        vFaces = swBody.GetFaces
        For j = 0 To UBound(vFaces)
            Set swFace = vFaces(j)
            Set swEnt = swFace
            Debug.Print swPart.SetEntityName(swEnt, "F" &
FaceCounter)
            FaceCounter = FaceCounter + 1
        Next j
    Next i

' 5. Selection of specified face
        Set swFace = swPart.GetEntityBy-
Name(UserForm1.TextBox2.Text, swSelFACES)
        Set swEnt = swFace
```

```vba
' 5b. Check for specified face
        If swEnt Is Nothing Then
        MsgBox "Face " & TextBox2.Text & " does not exist
or already have a hole _
        & vbNewLine & "Please choose another."
        Exit Sub
      Else
        swEnt.Select4 True, swSelectData
      End If

' 6. Sketch creation for hole
  swPart.SketchManager.InsertSketch True
  Set swSketch = swPart.SketchManager.CreateCircleByRa-
dius(x, y, z, Diameter)
  Set swFeature = swPart.FeatureManager.FeatureCut3(True,
False, False, 6, 0, Depth, 0, False, False, False, _
  False, 0, 0, False, False, False, False, False, True,
True, True, True, False, 0, 0, False)

End Sub
'-------------------------------------------------------
' End of code for Hole_Button_Click command button
'-------------------------------------------------------

' Exit form - end macro
Private Sub Exit_Button_Click()
 End
End Sub
' Form initialization and default values
Private Sub UserForm_Initialize()
 TextBox1.Text = "E1"
 TextBox2.Text = "F1"
 Diameter_TextBox.Text = 10
 Depth_TextBox.Text = 25
 x_TextBox.Text = 0
 y_TextBox.Text = 0
 z_TextBox.Text = 0
 Angle_TextBox.Text = 45
 Distance_TextBox.Text = 2
 Chamfer_Button = True
End Sub

'********************************************************
' End of program
'********************************************************
```

References

1. Babic B, Nesic N, Miljkovic Z (2008) A review of automated feature recognition with rule-based pattern recognition. Comput Ind 59:321–337. https://doi.org/10.1016/j.compind.2007.09.001
2. Shi Y, Zhang Y, Xia K, Harik R (2020) A critical review of feature recognition techniques. Comput Aided Des Appl 17:861–899. https://doi.org/10.14733/cadaps.2020.861-899
3. Venkataraman S, Sohoni M, Kulkarni V (2001) A graph-based framework for feature recognition. Proc Symp Solid Model Appl. https://doi.org/10.1145/376957.376980
4. Dalvi C, Rathod M, Patil S et al (2021) A survey of AI-based facial emotion recognition: features, ML DL techniques, age-wise datasets and future directions. IEEE Access 9:165806–165840. https://doi.org/10.1109/ACCESS.2021.3131733
5. Verma AK, Rajotia S (2010) A review of machining feature recognition methodologies. Int J Comput Integr Manuf 23:353–368. https://doi.org/10.1080/09511921003642121
6. Han J, Requicha AAG (1997) Integration of feature based design and feature recognition. Comput Des 29:393–403. https://doi.org/10.1016/S0010-4485(96)00079-6
7. Tzotzis A, Tsagaris A, Tapoglou N, Kyratsis P (2023) High-precision CAD-based simulation for turning considering tool microgeometry. Int J Mechatron Manuf Syst 16:83–95. https://doi.org/10.1504/IJMMS.2023.132023
8. Tzotzis A, Manavis A, Efkolidis N, Kyratsis P (2021) CAD-based automated G-code generation for drilling operations. Int J Mod Manuf Technol 13:177–184. https://doi.org/10.54684/ijmmt.2021.13.3.177
9. Sunil VB, Pande SS (2009) Automatic recognition of machining features using artificial neural networks. Int J Adv Manuf Technol 41:932–947. https://doi.org/10.1007/s00170-008-1536-z
10. Kyratsis P, Tzotzis A, Davim JP (2023) FEM-based study of AISI52100 steel machining : a combined 2D and 3D. In: 3D FEA simulations in machining, pp 47–62
11. Guo L, Zhou M, Lu Y et al (2021) A hybrid 3D feature recognition method based on rule and graph. Int J Comput Integr Manuf 34:257–281. https://doi.org/10.1080/0951192X.2020.1858507
12. Tsagaris A, Polychroniadis C, Tzotzis A, Kyratsis P (2024) Cost-effective robotic arm simulation and system verification. Int J Intell Syst Appl 16:1–12. https://doi.org/10.5815/ijisa.2024.02.01
13. Tapoglou N, Efstathiou C, Tzotzis A, Kyratsis P (2023) Study of the topography of face milled surfaces using CAD-based simulation. In: Kyratsis P, Manavis A, Davim JP (eds) Computational design and digital manufacturing. Springer International Publishing, Cham, pp 159–166
14. Manavis A, Tzotzis A, Tsagaris A, Kyratsis P (2022) A novel computational-based visual brand identity (CbVBI). Machines
15. Gella-Marín R, Tzotzis A, García-Hernández C, et al (2021) CAD software integration with programming tools for modelling, measurement and verification of surfaces. In: Experiments and simulations in advanced manufacturing. pp 91–116
16. Tzotzis A, Garcia-Hernandez C, Huertas-Talon J-L et al (2017) Engineering applications using CAD based application programming interface. In: MATEC web of conferences, pp 1–7
17. Tzotzis A, Efkolidis N, García-Hernández C, Kyratsis P (2024) CAD-based automated USV hull design. In: Guxho G, Kosova Spahiu T, Prifti V et al (eds) Proceedings of the joint international conference: 10th textile conference and 4th conference on engineering and entrepreneurship. Springer Nature Switzerland, Cham, pp 354–362
18. Kyratsis P, Tzotzis A, Manavis A (2021) Computational design and digital fabrication. In: Kumar S, Rajurkar KP (eds) Advances in manufacturing systems. Springer Singapore, Singapore, pp 1–16
19. Kyratsis P, Tzotzis A, Tzetzis D, Sapidis N (2018) Pneumatic cylinder design using cad-based programming. Acad J Manuf Eng 16:107–113

Chapter 4
Automated Assembly of Mechanical Systems: A Case Study

Abstract Assembling multiple identical components with a base part is common in industrial processes. Therefore, assembly automation within the graphical environment constitutes a critical aspect of a system's development stage. Present paper introduces the utilization of Computer-Aided Design (CAD)-based programming for the realization of automated assembly tasks. Traditional programming with the Visual Basic for Application (VBA™) language was employed for this purpose. Specifically, the API methods that are available in SolidWorks™ CAD system were utilized. The assembly process between a set of bolt-shaped components and an orthogonal solid was used to test the developed tool. API-based programming was employed to search for the appropriate geometric objects and handle the obtained data. The textual programming was preferred instead of the visual programming or the use of tools that involve Artificial Neural Networks (ANNs) and similar methods, for simplicity. Moreover, the traditional methods were used, since they require far less computational resources and do not require specialized tools at all. By combining the developed code with a graphical User Interface (UI), the tool becomes more convenient and practical, as well as more efficient.

Keywords Application programming interface (API) · Automated assembly · Computer-aided design (CAD) · CAD-based manufacturing · SolidWorks™ · User interface (UI) · Visual basic for applications (VBA™)

4.1 Introduction

In modern manufacturing, automated assembly has emerged as a crucial component, significantly enhancing production efficiency and product consistency. One of the key enablers of automated assembly is the integration of Computer-Aided Design (CAD) software with programming tools that guide assembly processes. By leveraging CAD software, manufacturers can create highly detailed models of components and assemblies, which can then be translated into instructions for automated machinery [1]. This fusion of CAD software with automation technology allows for the seamless translation of design intent into physical products, minimizing human intervention and reducing errors. CAD software has traditionally been used for creating detailed digital representations of parts and assemblies. These digital models serve as blueprints for manufacturing processes. However, with the advent of automated assembly, the role of CAD software has expanded significantly.

Automated assembly relies heavily on the precision and detail offered by CAD models. The geometrical and functional data embedded in these models is used to program robots and other automated machinery. This process, known as CAD-based programming, involves the extraction of assembly instructions directly from CAD models. These instructions include details such as the sequence of assembly, the orientation of parts, and the specific operations required at each step. For instance, a study by Gu and Yan [2] demonstrated how CAD models can be utilized to generate robotic assembly instructions automatically. Their research highlighted the efficiency of using CAD-based programming to eliminate the need for manual coding, thus speeding up the programming process and reducing errors associated with manual interventions. Similar studies [3, 4] prove the efficiency of virtual assembly environments.

CAD-based automated assembly significantly reduces the time required for production. By directly using CAD models for programming, the need for extensive manual input is minimized, allowing for faster setup times. This efficiency is particularly beneficial in high-volume production environments where speed and consistency are critical. A study by Kreis et al. [5] emphasized the role of CAD software in streamlining the assembly process, noting significant reduction in cycle time when CAD-based automation was implemented in a case study involving automotive parts assembly. Automated assembly systems that utilize CAD programming ensure that each product is assembled with the same level of precision. The digital models in CAD provide exact specifications for parts, ensuring that the assembly process adheres to these specifications consistently. This precision reduces variability in the final products, leading to higher quality and fewer defects [6, 7].

CAD-based automated assembly systems offer a high degree of flexibility as shown in the study by Viganò and Osorio-Gómez [8]. Manufacturers can easily modify CAD models and update the corresponding assembly instructions without significant downtime. This capability is particularly useful in industries where product designs frequently change or where customization is important. Research by Corallo et al. [9] discussed the adaptability of CAD-based automation in the aerospace industry, where frequent design changes are common. The study found that CAD-integrated assembly systems could accommodate these changes with minimal disruption to the production line. By streamlining the assembly process and reducing the need for manual labor, CAD-based automation can lead to significant cost savings [10]. The initial investment in automation technology is often offset by the long-term savings in labor, material waste, and rework costs due to errors.

Despite the numerous advantages, there are challenges associated with implementing CAD-based automated assembly systems. Integrating CAD software with automated assembly systems requires significant technical expertise. Moreover, the transition from traditional manual assembly to automated systems can be complex and requires careful planning and execution. Additionally, there may be costs associated with training personnel to operate and maintain these systems. According to a study by Hammar and Norström [11], the high upfront costs of CAD-based automation can be a barrier for small and medium-sized enterprises. The study suggested that while the long-term benefits are significant, the initial financial outlay may deter some businesses from adopting this technology. Automated assembly systems require regular maintenance to ensure they operate efficiently. Additionally, as CAD software evolves, updates may be necessary to maintain compatibility with the assembly systems. These updates can be costly and may require downtime, affecting production schedules. With the increasing reliance on digital models and automated systems, data management becomes a critical issue. Ensuring that CAD data is accurately maintained and protected from unauthorized access is essential. The integration of CAD systems with the Internet of Things (IoT) and cloud computing adds another layer of complexity in terms of data security.

The future of CAD-based automated assembly looks promising, driven by advancements in technology such as Artificial Intelligence (AI), machine learning, and IoT. These technologies are expected to further enhance the capabilities of CAD systems, making automated assembly even more efficient and adaptable. AI and machine learning are expected to play a significant role in the future of automated assembly. These technologies can be integrated with CAD systems to optimize assembly processes in real-time. For example, AI algorithms can analyze data from previous assemblies to identify potential improvements in the process, thereby increasing efficiency and reducing errors. A study by Salchner et al. [12] explored the potential of AI-driven CAD-based assembly systems. The research indicated that AI could enhance the adaptability of automated systems, enabling them to handle more complex tasks with minimal human intervention. The integration of IoT and cloud computing with CAD-based automated assembly systems is another area of significant potential. IoT devices can provide real-time data on the status of assembly lines, allowing for more precise control and monitoring. Cloud computing, on the other hand, offers scalable solutions for storing and processing the vast amounts of data generated by these systems. As discussed in the studied [13, 14], the combination of IoT, cloud computing, and CAD software could revolutionize automated assembly, enabling real-time collaboration and data sharing across different locations and production facilities.

Finally, the future of manufacturing is moving toward mass personalization, where products are tailored to individual customer specifications. CAD-based automated assembly systems are well-suited to this trend, as they can quickly adapt to design changes and produce customized products efficiently. Recent researches [15, 16] suggest that CAD-based automation will be instrumental in enabling manufacturers to meet the growing demand for personalized products. Their study highlighted how CAD-integrated systems could facilitate the production of customized items at scale without compromising on efficiency.

The present work presents the applicability of API programming [17] in assembly, with an aim to automate the assembly process through CAD systems. In specific, the automated assembly of an orthogonal plate with a set of cylindrical-shaped components, similar to the assembly of multiple in-line bolts or rivets with a mechanical part containing holes. The process is facilitated by the developed application tool, without the need for specialized software or programming language. Using the API methods of SolidWorks™ for traversing and identifying the components' geometric features [18–20], the automated assembly process is achieved. This work, demonstrates the abilities of traditional programming within the CAD environment, without utilizing AI methods or similar advanced techniques.

4.2   Material and Methods

4.2.1   User Interface

Figure 4.1 illustrates the User Interface (UI) designed for automating the assembly process within a CAD environment. The display area shows a preview of the assembly. The preview typically includes a visual representation of the components that will be assembled. In this instance, the preview shows a series of cylindrical components assembled on a rectangular base, giving the user a quick visual confirmation of the assembly setup before proceeding. The instance constitutes a simplified version of the assembly between a mechanical component containing a series of machined holes and the equivalent set of bolts or rivets. The textbox serves as an input field for pointing the location of the files involved in the assembly. It directs the user to input or confirm the directory path from which the assembly files will be loaded or saved. In the example provided, the path "C:\Test_Folder" is shown, indicating the directory where the assembly files are located. The "Assembly" command button is prominently placed at the bottom left of the UI. When clicked, it triggers the automated assembly process using the specified files located in the path provided in the input field labeled "File Path". The button is labeled "Assembly" clearly indicating its function. Positioned to the right of the "Assembly" button, the "Exit" button allows the user to close the tool when they have completed their task, providing a straightforward way to exit the application.

This UI is designed to streamline the automated assembly process in a CAD environment [7, 21]. It gives users a visual confirmation of what the final assembly will look like, reducing the risk of errors. Moreover, it allows users to specify the location of the necessary files, making the assembly process more efficient. Together, these elements provide a user-friendly interface for managing and executing automated assemblies of the specified format (holed component—set of bolts) with minimal manual intervention.

Fig. 4.1 UI of the
automated assembly tool

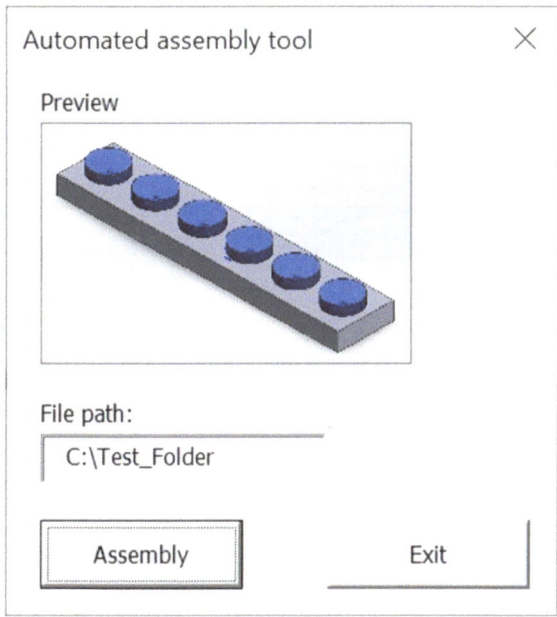

4.2.2 Assembly Process Flow

Figure 4.2 outlines the steps involved in assembling different components into a complete assembly within a CAD environment. First step involves defining the necessary variables or geometric objects (such as faces, edges, or points) that will be used throughout the assembly process. Next step, ensures that the correct model document is activated and ready for modification. The code identifies and retrieves specific faces of the 3D model. These faces are likely to be critical surfaces that will interact with other components during assembly. Specifically, they contain circular edges that are necessary for the mate between the holes and the bolt-shaped items. The system retrieves the circular edges from the model's faces in a similar manner found in the study by Tzotzis et al. [22]. Circular edges are often used for aligning or mating parts, such as fitting a bolt into a hole, as already mentioned previously. The retrieved circular edges are grouped into a collection. This collection can be used later to apply consistent operations or constraints across all the circular edges. The sixth step involves traversing through the collection of geometric objects (faces and circular edges) created earlier. Safe entities, such as constraints or reference points, are created to ensure the assembly process can proceed without errors. Next, the specific part document to be added to the assembly is opened. This part represents the bolt or rivet. The opened part is added to the assembly document, which is the document where multiple parts come together to form a complete product or model. The system revisits the collection of edges and creates mates. Mates are constraints that define how parts fit together, such as aligning a pin with a hole, which is the

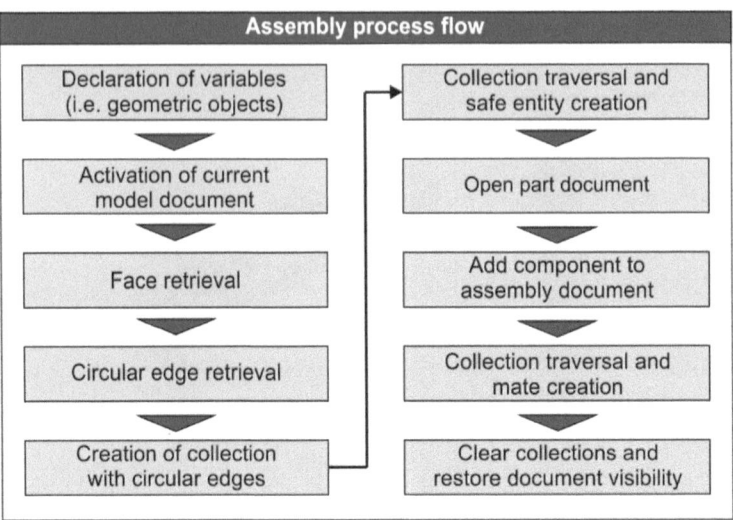

Fig. 4.2 Process flow during the assembly

case, or ensuring two faces are flush. Once the assembly is complete, the collections of geometric objects are cleared, and the document's visibility settings are restored. This step ensures that the model is clean and ready for further analysis or operations. Moreover, any new documents that are opened can become visible. It should be noted that prior to the automated assembly process, any components that involve in the assembly process are opened silently, so that visual nuisance is avoided. Moreover, this technique ensures that computer resources such as memory, are sufficient.

The code focuses on preparing the model by retrieving relevant faces and circular edges, as well as organizing them into collections for easy reference. In addition, it deals with the actual assembly process: opening part documents, adding components to the assembly, and using the collections to efficiently create mates between the parts. After the necessary relationships between components are established, the collections are cleared, and the assembly document is finalized. This assembly process flow is designed to streamline the creation of assemblies in CAD software, ensuring that parts are correctly aligned and mated, and that the process is both efficient and error-free.

4.2.3 Process Code

Figure 4.3 illustrates the process diagram. The process begins with selecting a geometric object (step 1), which refers to a specific geometric feature or element within the model. Specifically, the user must manually select a face in the model that contains holes. This step involves identifying the surface where components, like screws, bolts, or rivets, will be assembled. Next, the first decision point (condition A) in the workflow appears. It checks whether the selected geometric object (GO) is a face. If it is, the workflow proceeds to step 2. If not, the workflow loops back to the "Topology Object Selection" step (step 1) for another selection. If this is the case, a message prompts the user to select an appropriate face. In the case where the selected topology object is a face, the workflow creates a "safe entity" from this face (step 2). A safe entity could be a locked or referenced version of the face that can be used without modifying the original model. The code then traverses the edges of the selected face (step 3). This step involves identifying and recording the edges associated with the face. After traversing the edges, the code creates a collection of circular edges (step 4). This collection will be used for tasks like identifying where the fasteners can be placed. The code then traverses through the created collection of circular edges, examining each one for further processing (step 5). Condition B, is the second decision point in the workflow. It checks whether the circular edge (CE) belongs to a circular surface (CS). If it is, the workflow proceeds to step 6. If no, the workflow loops back to traverse the collection again. If the circular edge belongs to a circular surface, a safe entity is created for these circular surfaces, similar to the face entity created in step 2. Next, the part document containing the components to be assembled is opened (step 7). Thereafter, the required component is added to the assembly at the selected location (step 8). Mates are constraints applied between components to position them relative to each other in the assembly. In this case, the first mate (concentric) is responsible for axially aligning each hole with one copy of the component and the second one (coincident), takes care the placement of the components on the drilled surface of the plate (step 9). Eventually, the code clears the collections used during the assembly process and restores the model settings to their original state (step 10). Specifically, it reenables the visibility of the new documents, since it was disabled just before they were opened for the assembly. This technique is common when a large number of parts are involved in the assembly.

The most important API methods for the development of the code are presented in Table 4.1. "GetSelectedObjectType3" determines the type of the currently selected object in the SolidWorks™ environment and helps in identifying what kind of entity (e.g., face, edge, vertex) is selected, allowing the program to handle it appropriately depending on its type. "GetSelectedObject6" method retrieves the object that is currently selected and provides a direct reference to the selected entity, which can then be used for further operations such as analysis or modification. "GetSafeEntity" obtains a reference to a "safe" entity, which allows you to interact with it without directly altering the original model. Moreover, it ensures that operations on the entity, such as transformations or measurements, do not accidentally modify or corrupt the original geometry in the model. "GetFirstLoop" retrieves the first loop on a face, where a loop is a continuous path of connected edges. It is useful in operations involving face loops, such as determining the boundary of a hole or an opening in a surface. This first loop might not always be the outer boundary but could represent an inner loop or hole. "GetEdges" collects all the edges within a specified loop. This method facilitates operations where working with all the edges of a loop is required,

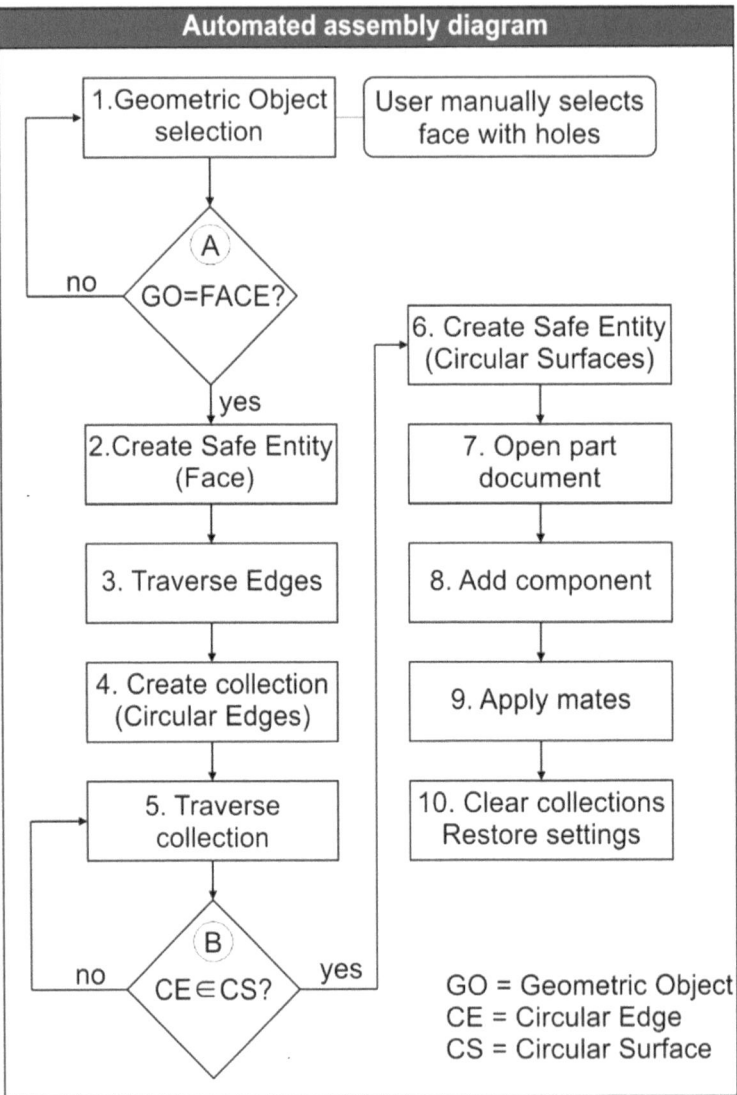

Fig. 4.3 Process diagram for the automated assembly

such as generating profiles for features or identifying the outline of a loop for further processing. "GetCurve" retrieves the underlying geometric curve associated with a specific edge. Thus, it allows access to the mathematical representation of an edge, which can be used for precise calculations, creating new geometry, or understanding the shape and continuity of the edge. "GetNext" moves to the next loop on a face after the current one and is useful in iterating through multiple loops on a face, such as when dealing with complex faces that have multiple holes or boundaries.

"GetTwoAdjacentFaces2" retrieves the two faces that are adjacent to a specified edge. It is important for operations where the relationship between faces is critical, such as in creating fillets, chamfers, or determining how two surfaces meet. "GetSurface" retrieves the surface associated with a selected face and is used for accessing the surface geometry, since it is crucial for operations like surface analysis, creating new features based on surface curvature, or understanding the topological properties of the model.

"OpenDoc6" simply opens an existing document (such as a part or assembly file) in SolidWorks™ and returns a reference to the document. "AddComponent4" adds a specified component to the assembly with specified configuration options. Essentially, it automates the process of adding components to an assembly, allowing for controlled placement and configuration, which is crucial in large assemblies or automated assembly generation. "SelectByID2", as the name suggests, selects an entity (such as a face, edge, or component) by its ID in the model. It greatly facilitates precise selection of entities within the model, enabling targeted operations like applying mates, making measurements, or modifying specific parts of the assembly. Finally, the "CreateMate" method creates a mate (a geometric constraint) between two components using the specified feature data object, defining how components are positioned relative to each other in an assembly, ensuring correct alignment, orientation, and movement constraints as per design requirements.

These methods collectively enable the automation of various tasks in the assembly process, reducing manual input and increasing efficiency in design workflows. The complete code for the process is available at the appendix section. For the code to

Table 4.1 Basic API methods for the automated assembly process

Method	Explanation
GetSelectedObjectType3	Gets the type of object selected
GetSelectedObject6	Gets the selected object
GetSafeEntity	Gets a safe entity
GetFirstLoop	Gets the first loop in this face, which is not necessarily the outer loop
GetEdges	Gets all the edges in the loop
GetCurve	Gets the underlying curve for this edge
GetNext	Gets the next loop on the face
GetTwoAdjacentFaces2	Gets the two faces adjacent to an edge
GetSurface	Gets the surface referenced by this face
OpenDoc6	Opens an existing document and returns a pointer to the document object
AddComponent4	Adds the specified component for the specified configuration options to this assembly
SelectByID2	Selects the specified entity
CreateMate	Creates a mate with the specified feature data object

function, the appropriate form (see Fig. 4.1) should be created and the corresponding elements must be placed and named accordingly.

4.3 Results

Figure 4.4 illustrates the assembly process, which involves five key stages, from selecting features on the plate to positioning and mating the fasteners within those features. Stage 1 involves the manual selection of the face of the plate that contains the holes. This is done by the user in the CAD software interface, where the face of interest is clicked and highlighted. This stage is crucial as it identifies the specific area of the plate where the fasteners will be placed and mated. Other than that, the user has no intervention. After the face with the holes has been selected, the tool automatically identifies and selects the circular edges of these holes (stage 2). Circular edges are important for mating operations, especially when dealing with cylindrical components such as bolts and rivets, as they help ensure proper alignment and positioning. Following the selection of circular edges, the tool proceeds to automatically select the circular faces associated with these edges (stage 3). These faces are typically the inner surfaces of the holes. The selection of these faces is necessary for accurate placement of the fasteners, ensuring that they fit perfectly within the holes. In stage 4, the bolt-like components are introduced into the assembly. The tool positions these components relative to the selected circular faces and edges from the previous stages. Proper positioning is essential for the subsequent mating process, as it ensures that the components are aligned correctly with the holes on the plate. Stage 5 involves the creation of mates between the fasteners and the plate. A mate in CAD terms is a constraint that defines how two components interact with each other, such as making them concentric or coincident. Here, the fasteners are mated with the circular holes in the plate, ensuring that each fastener is properly seated and aligned. This stage finalizes the assembly, making the fasteners a part of the overall model.

4.4 Conclusion

Automated assembly based on CAD software programming represents a significant advancement in manufacturing, offering numerous benefits such as increased efficiency, accuracy, and flexibility. However, the successful implementation of these systems requires careful consideration of the associated challenges, including integration complexity, high initial costs, and data management issues. As technology continues to evolve, the integration of AI, IoT, and cloud computing with CAD-based automated assembly systems promises to further enhance their capabilities, paving the way for more innovative and efficient manufacturing processes.

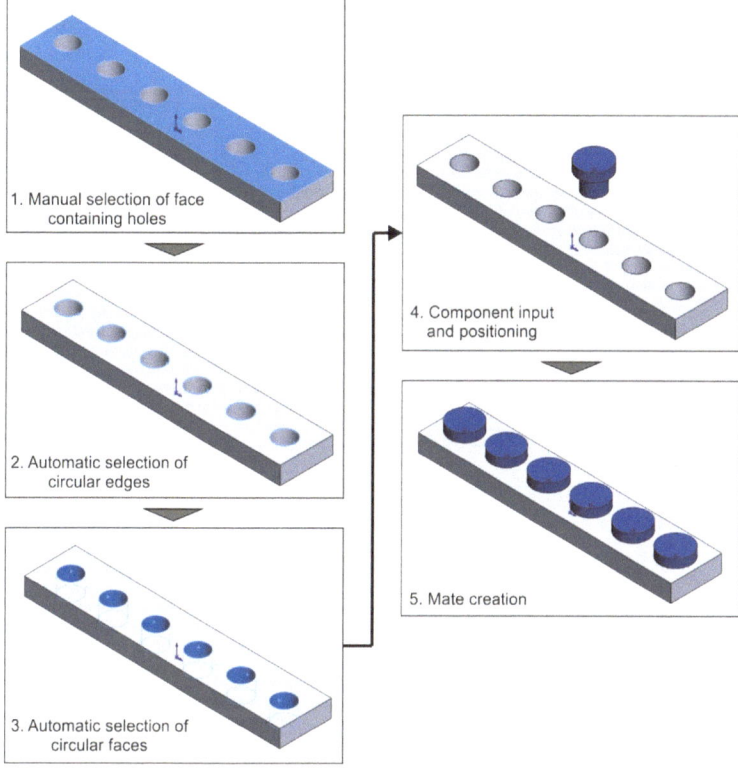

Fig. 4.4 Assembly process of the plate with the components

This work demonstrated the use of traditional programming, within the API environment of a commercially available CAD system, for the assembly process automation. The paradigm presented represents a typical process in the manufacturing industry, the assembly between a base part and a series of fastener components. Finally, the next conclusions can be drawn for the use of this method:

- The computational resources required for both development and application are relatively low.
- The initial cost is relatively low. Especially when compared to projects that involve IoT and AI, since in this case the only resource required is the CAD software.
- It is possible to expand the usability of the tool to other types of assembly.
- The assembly process concludes promptly, with minor effort from the user.
- No special training is required for the operation of the tool, whereas the development and editing of a similar tool would require programming knowledge of intermediate level.

Appendix

```
'***********************************************************
' Automated assembly of bolts/rivets - holes by creating
' geometric object collection
'***********************************************************

Option Explicit

Dim swApp As SldWorks.SldWorks
Dim swModel As SldWorks.ModelDoc2
Dim swAssy As SldWorks.AssemblyDoc
Dim swSelMgr As SldWorks.SelectionMgr
Dim swFace As SldWorks.Face2
Dim swSelFaceEnt As SldWorks.Entity

Dim collCircEdges As New Collection
Dim collCylFaces As New Collection
Dim collCylSafeFaces As New Collection

Dim i As Integer
Dim errors As Long
Dim warnings As Long
Dim boolstatus As Boolean
Dim Path As String
Dim CompPath As String

'-----------------------------------------------------------
' Start of code for Assembly command button
'-----------------------------------------------------------
Private Sub Assembly_Button_Click()

' 1. Connection to SolidWorks
 Set swApp = Application.SldWorks
 Set swModel = swApp.ActiveDoc
 Set swAssy = swModel
 Path = Assembly_Form.Path_TextBox.Text
 CompPath = Path & "\Pin.sldprt"

' 2. Check if assembly document is open
 If swModel Is Nothing Then
  MsgBox ("Please open an assembly document.")
  Exit Sub
 End If

' 3. Subroutines execution
 Call GetFace           'Gets the face the user selects
```

```
  swAssy.EditAssembly      'Checks if the assembly is edited
  Call GetCircularEdges    'Checks for cylindrical edges
  Call GetCylFaces         'Checks for cylindrical faces
  Call OpenCompModel       'Opens components
  Call AddComponents       'Inserts Pin.sldprt
  Call Finalize            'Restores initial settings

End Sub
'--------------------------------------------------------
' End of code for Assembly command button
'--------------------------------------------------------
Private Sub GetFace()

    Set swSelMgr = swModel.SelectionManager
' Checks the selected geometric object, if it is a face
then it creates a safe entity
    If swSelMgr.GetSelectedObjectType3(1, -1) = swSe-
lectType_e.swSelFACES Then
        Set swFace = swSelMgr.GetSelectedObject6(1, -1)
        Set swSelFaceEnt = swFace
        Set swSelFaceEnt = swSelFaceEnt.GetSafeEntity
' Display message to prompt user select appropriate face
    Else
        swApp.SendMsgToUser2 "Select face that contains
holes." swMbWarning, swMbOk
        End
    End If
End Sub

Private Sub GetCircularEdges()
    Dim swLoop As SldWorks.Loop2
    Dim swEdges As Variant
    Dim swCurve As SldWorks.Curve

    Set swFace = swSelFaceEnt

    Set swLoop = swFace.GetFirstLoop
' Loop for traversing edges and creating collection with
circular edges
    While Not swLoop Is Nothing
        If Not swLoop.IsOuter Then
            swEdges = swLoop.GetEdges
            For i = 0 To UBound(swEdges)
                Set swCurve = swEdges(i).GetCurve
                If swCurve.IsCircle Then
                 collCircEdges.Add swEdges(i)
                End If
            Next i
```

```
        End If
        Set swLoop = swLoop.GetNext
    Wend
End Sub

Private Sub GetCylFaces()
    Dim vFaces As Variant
    Dim swSurface1 As SldWorks.Surface
    Dim swSurface2 As SldWorks.Surface
    Dim swCylFaceEnt As SldWorks.Entity

' Loop for traversing the collection and creating safe
entities
    For i = 1 To collCircEdges.Count
        vFaces = collCircEdges.Item(i).GetTwoAdja-
centFaces2
        Set swSurface1 = vFaces(0).GetSurface
        Set swSurface2 = vFaces(1).GetSurface
        If swSurface1.IsCylinder Then
            Set swCylFaceEnt = vFaces(0)
            Set swCylFaceEnt = swCylFaceEnt.GetSafeEntity
            collCylFaces.Add swCylFaceEnt
        ElseIf swSurface2.IsCylinder Then
            Set swCylFaceEnt = vFaces(1)
            Set swCylFaceEnt = swCylFaceEnt.GetSafeEntity
            collCylFaces.Add swCylFaceEnt
        End If
    Next i
End Sub

' Open Pin.sldprt
Private Sub OpenCompModel()
    swApp.DocumentVisible False, swDocPART
    swApp.OpenDoc6 CompPath, 1, 0, "", errors, warnings
End Sub

Private Sub AddComponents()
    Dim swComp As SldWorks.Component2
    Dim CompName As String
    Dim SelData As SldWorks.SelectData
    Dim strings As Variant
    Dim AssyName As String

    strings = Split(swModel.GetTitle, ".")
    AssyName = strings(0)

' Loop for traversing the collection and creation of ma-
tes
```

```
  For i = 1 To collCircEdges.Count
  Set swComp = swAssy.AddComponent4(CompPath, "", 0, 0, 0)
       CompName = swComp.Name2
swSelFaceEnt.Select4 False, SelData
swModel.Extension.SelectByID2 "Front Plane@" & CompName _
& "@" & AssyName, "PLANE", 0, 0, 0, True, 1, Nothing, 0
swAssy.CreateMate swMateCOINCIDENT, swMateAlignCLOSEST, _
False, 0, 0, 0, 0, 0, 0, 0, 0, False, errors

        collCylFaces.Item(i).Select4 False, SelData
        swModel.Extension.SelectByID2 _
           "Point1@Origin" & "@" & CompName & "@" &
AssyName, "EXTSKETCHPOINT", 0, 0, 0, True, 1, Nothing, 0
swAssy.CreateMate swMateCONCENTRIC, swMateAlignCLOSEST, _
False, 0, 0, 0, 0, 0, 0, 0, 0, False, errors
    Next i
End Sub

' Clean collections - restore initial settings
Private Sub Finalize()
    Set collCircEdges = Nothing
    Set collCylFaces = Nothing
    swApp.DocumentVisible True, swDocPART
    swModel.ClearSelection2 True
End Sub

' Exit form
Private Sub Exit_Button_Click()
 End
End Sub

' Component file path
Private Sub UserForm_Initialize()
 Path_TextBox = "C:\Test_Folder"
End Sub
```

References

1. Tzotzis A, Garcia-Hernandez C, Huertas-Talon J-L et al (2017) Engineering applications using CAD based application programming interface. In: MATEC web of conferences, pp 1–7
2. Gu P, Yan X (1995) CAD-directed automatic assembly sequence planning. Int J Prod Res 33:3069–3100.https://doi.org/10.1080/00207549508904862
3. Leu MC, ElMaraghy HA, Nee AYC et al (2013) CAD model based virtual assembly simulation, planning and training. CIRP Ann 62:799–822. https://doi.org/10.1016/j.cirp.2013.05.005
4. Lin YJ, Farahati R (2003) CAD-based virtual assembly prototyping—a case study. Int J Adv Manuf Technol 21:263–274. https://doi.org/10.1007/s001700300031
5. Kreis A, Hirz M, Rossbacher P (2020) Cad-automation in automotive development—potentials, limits and challenges. Comput Aided Des Appl 18:849–863. https://doi.org/10.14733/cadaps. 2021.849-863
6. Kyratsis P, Gabis E, Tzotzis A et al (2019) CAD based product design: a case study. Int J Mod Manuf Technol 11:88–93

7. Kyratsis P, Tzotzis A, Tzetzis D, Sapidis N (2018) Pneumatic cylinder design using cad-based programming. Acad J Manuf Eng 16:107–113

8. Viganò R, Osorio-Gómez G (2012) Assembly planning with automated retrieval of assembly sequences from CAD model information. Assem Autom 32:347–360. https://doi.org/10.1108/01445151211262410

9. Corallo A, Laubacher R, Margherita A, Turrisi G (2009) Enhancing product development through knowledge-based engineering (KBE). J Manuf Technol Manag 20:1070–1083. https://doi.org/10.1108/17410380910997218

10. Manavis A, Tzotzis A, Tsagaris A, Kyratsis P (2022) A novel computational-based visual brand identity (CbVBI). Machines

11. Hammar F, Norström R (2020) An investigation into barriers for flexible automation in SMEs an investigation into barriers for flexible automation in SMEs

12. Salchner M, Stadler S, Hirz M et al (2016) Multi-CAD approach for knowledge-based design methods. Comput Aided Des Appl 13:471–483. https://doi.org/10.1080/16864360.2015.1131540

13. Reddy EJ, Reddy KS, Reddy PN (2024) Recent advancements in knowledge-based parametric modeling of mechanical components. In: Yadav S, Arora PK, Sharma AK, Kumar H (eds) Proceedings of third international conference in mechanical and energy technology. Springer Nature Singapore, Singapore, pp 401–411

14. Tariki K, Kiyokawa T, Nagatani T et al (2021) Generating complex assembly sequences from 3D CAD models considering insertion relations. Adv Robot 35:337–348. https://doi.org/10.1080/01691864.2020.1863258

15. Yuan Y, Huh JH (2018) Customized CAD modeling and design of production process for one-person one-clothing mass production system. Electron. https://doi.org/10.3390/electronics7110270

16. Liu F (2020) Fast industrial product design method and its application based on 3D CAD system. Comput Aided Des Appl 18:118–128. https://doi.org/10.14733/cadaps.2021.S3.118-128

17. Kyratsis P, Tzotzis A, Manavis A (2021) Computational design and digital fabrication. In: Kumar S, Rajurkar KP (eds) Advances in manufacturing systems. Springer Singapore, Singapore, pp 1–16

18. Tzotzis A, Tsagaris A, Tapoglou N, Kyratsis P (2023) High-precision CAD-based simulation for turning considering tool microgeometry. Int J Mechatron Manuf Syst 16:83–95. https://doi.org/10.1504/IJMMS.2023.132023

19. Gella-Marín R, Tzotzis A, García-Hernández C, et al (2021) CAD software integration with programming tools for modelling, measurement and verification of surfaces. In: Experiments and simulations in advanced manufacturing, pp 91–116

20. Tapoglou N, Efstathiou C, Tzotzis A, Kyratsis P (2023) Study of the topography of face milled surfaces using CAD-based simulation. In: Kyratsis P, Manavis A, Davim JP (eds) Computational design and digital manufacturing. Springer International Publishing, Cham, pp 159–166

21. Tzotzis A, Efkolidis N, García-Hernández C, Kyratsis P (2024) CAD-based automated USV hull design. In: Guxho G, Kosova Spahiu T, Prifti V et al (eds) Proceedings of the joint international conference: 10th textile conference and 4th conference on engineering and entrepreneurship. Springer Nature Switzerland, Cham, pp 354–362

22. Tzotzis A, Manavis A, Efkolidis N, Kyratsis P (2021) CAD-based automated G-code generation for drilling operations. Int J Mod Manuf Technol 13:177–184. https://doi.org/10.54684/ijmmt.2021.13.3.177

Chapter 5
Integration of CAD Software Programming for G-code Generation

Abstract Nowadays, automated G-code generation can be achieved with either free software that are available on the web or with specialized Computer-Aided Manufacturing (CAM) systems. In the first case, 3D parts cannot be directly handled, whereas in the second one, the purchase cost is usually very high. The present paper introduces a way of utilizing 3D Computer-Aided Design (CAD) systems to generate G-code for drilling operations. The idea is to supplement free CAM systems, enabling 3D part handling, without the need for specialized CAM software. The methods and functions available in the Application Programming Interface (API) of SolidWorks™ were combined to develop the code using the Visual Basic for Applications (VBA™) language. Specifically, the appropriate API methods were employed to identify the design features of the part, enabling the geometric parameters extraction. The aim of the tool is the Computer Numerical Control (CNC) code generation for the identified hole features or geometric surfaces that belong to holes. A User Interface (UI) was designed to allow the user to interact. Additionally, an embedded calculator was added for the automatic calculation of the cutting parameters (cutting speed, feed, and tool), with respect to the inputs.

Keywords Application programming interface (API) · Computer numerical control (CNC) · Computer-aided design (CAD) · CAD-based manufacturing · G-code · SolidWorks™ · Visual basic for applications (VBA)

5.1 Introduction

Computer-Aided Design (CAD) has become an indispensable tool in modern manufacturing, revolutionizing the way products are designed, prototyped, and produced. By leveraging CAD technology, manufacturers can achieve unprecedented levels of precision, efficiency, and innovation. CAD-based manufacturing integrates computer-aided design tools with manufacturing processes to streamline the production of complex and precise components. This integration has significantly reduced the time and cost associated with product development, while enhancing design accuracy and flexibility [1]. One of the key advantages of CAD-based manufacturing is its ability to facilitate rapid prototyping. By quickly converting digital designs into physical prototypes using technologies such as 3D printing, manufacturers can test and refine their products at a fraction of the traditional cost and time [2]. Computer-Aided Manufacturing (CAM) systems work in tandem with CAD software to automate the manufacturing process. This integration ensures seamless transitions from design to production, reducing lead times and improving overall efficiency [3]. CAM systems interpret CAD models to generate precise instructions for CNC machines and other automated equipment. The impact of CAD-based manufacturing is evident across numerous industries, including automotive, aerospace, healthcare, and consumer electronics. In the automotive industry, CAD tools have enabled the design of safer and more fuel-efficient vehicles [4]. In aerospace, CAD has facilitated the development of lightweight yet durable components, contributing to advancements in aircraft performance [5]. CAD-based manufacturing also enhances collaboration among multidisciplinary teams. Cloud-based CAD platforms allow engineers, designers, and stakeholders to access and modify designs in real time, regardless of their location. This collaborative approach accelerates decision-making and improves the overall quality of the final product [6]. Sustainability is a critical concern in modern manufacturing. CAD tools contribute to sustainable practices by optimizing material usage and minimizing waste. Through precise simulations and analysis, manufacturers can develop products that are not only high-quality but also environmentally friendly [7]. Despite its advantages, CAD-based manufacturing also presents challenges. The high cost of advanced CAD software and the need for skilled personnel to operate these systems can be barriers for small and medium-sized enterprises (SMEs). Additionally, ensuring data security and managing intellectual property rights in collaborative environments are critical considerations [8]. CAD-based manufacturing has fundamentally transformed the landscape of modern industry, offering significant benefits in terms of precision, efficiency, and innovation. As technologies continue to evolve, the integration of CAD with advanced manufacturing techniques will further enhance the capabilities of industries worldwide.

Studies related to manufacturing take advantage of the CAD integration. Tzotzis et al. [9] developed a CAD-based tool for simulating the cutting occurred during the turning process. Moreover, the tool takes into consideration the full geometry of the cutting tool involved in the process. Chacón et al. [10] presented a software tool that is able to convert 2D black and white images into a manufacturing file. In addition, the topology optimization integration into CAD/CAM systems is feasible. Tapoglou et al. [11] presented a simulation model for the face-milling cutting conditions effect evaluation, with respect to the produced surface quality of machined parts. The results of the model include the produced surface topography as well as surface roughness data. Similar works [12–14] deal with the design of high-accuracy cutting tools, with the aid of CAD-based programming and computational tools, that can be imported in Finite Element Analysis (FEA) [15], as well as with the integration of CAD with programming tools, for the modeling, measurement, and verification of spur gear surfaces. Computational design and robotic simulations is another topic [16–18] which benefits from the integration of CAD-based programming.

This study presents the development of a CAD API application for the automated recognition of machined holes on a part. The aim is to generate the equivalent machining code for drilling operations without the need for specialized software. Additionally, an effort was made to take into consideration typical drilling operations and their corresponding strategies. Finally, the code was written in the VBA™ programming language.

5.2   Material and Methods

5.2.1   User Interface

The User Interface (UI) was developed in such a way to prioritize simplicity and user-friendliness [19, 20]. It is divided into three sections. The first one relates to the machining parameters input and the material selection with the aid of the appropriate textboxes and listbox. It is noted that the listbox for the material selection enables the automated parameter definition according to the ISO standard for the workpiece materials. For example, the colored symbols P, M, K, N, S, and H correspond to the steels group, the stainless steels, the cast irons, the nonferrous alloys, the high-temperature alloys, and the hardened steels. The cutting conditions are then automatically set based on standardized values recommended by renowned cutting-tool manufacturers. The second section contains the preview window for displaying the code after the process completes, allowing the users to review it.

Finally, the third section, is responsible for generating the CNC code for the drilled part. It includes a textbox for entering the file saving path, where a text document with the code lines is saved. Moreover, a window displays the drilling operations involved depending on the type of the holes that are available on the machined part. The drilling operations are automatically selected by the application after recognizing the hole types. The two command buttons, "Generate Machining Data" and "Generate Machine Code", are responsible for the hole features identification process initiation for the attribute's designation and the generation of the machining respectively. The machine code is generated based on three key factors. The first one is the selected machining parameters, the second one is the number and type of the detected holes, and the third one is the chosen by the user drilling strategy. Figure 5.1 depicts the UI for the developed tool, showing the input fields, preview windows, and command buttons.

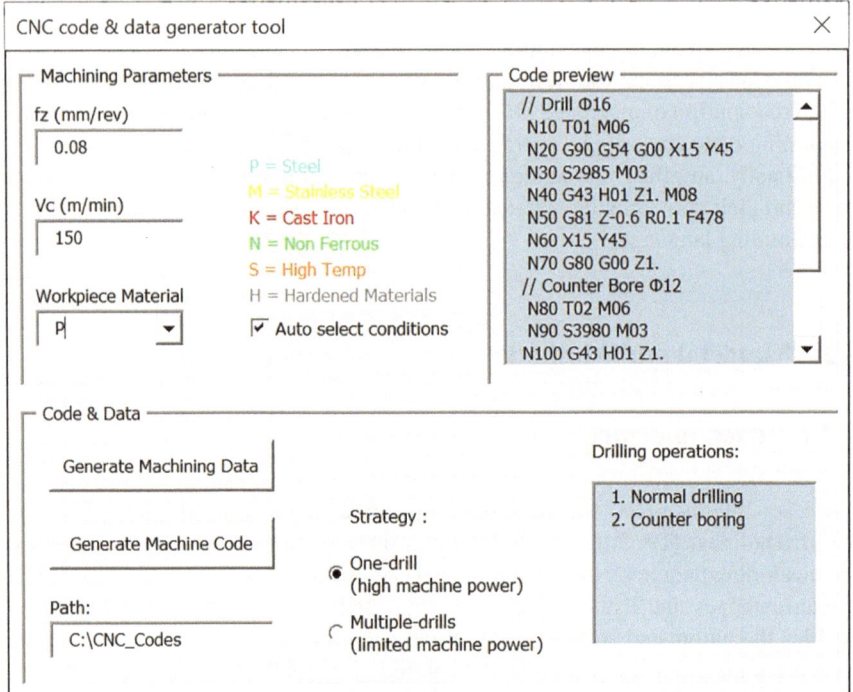

Fig. 5.1 The UI of the CNC code generator tool

5.2.2 Machining Code Generation Flow

The developed tool is able to automatically identify the design features related to holes of a part and the creation of attribute definitions in order to generate the G-code. Figure 5.2 is illustrating the whole process in steps. First of all, as in all cases, the variables related to the cutting speed, feed rate, hole position, diameter and depth, machine tool type, workpiece material group, attribute parameters, and geometric objects must be declared. Next step is the matching of the variables to the process options that are available to the UI. For instance, the listbox labeled "Workpiece Material" is connected to the corresponding material variable that is used in a conditional structure (if-then-else) for the machining parameters recommendation. It is noted that not all variables are matched with entities in the UI, many variables are matched with objects inside the code. Next, an already open part document is accessed so that the identification process can begin. The part should contain features related to cylindrical surfaces created either by a normally designed hole or by an extruded-cut feature. Next, the attribute definitions for the aforementioned parameters such as speed, feed, hole position in terms of coordinates, depth, and diameter are created. Following, the available solid bodies being identified in the open document, are used for the cylindrical surfaces' identification process, which then are being assigned to a unique attribute. The attributes are being displayed in the form of a graphical callout on the model. Thus, when the use presses the button "Generate Machining Data" the result is the immediate display of these callouts for each hole.

By considering the first six steps as the part of the code related to the attribute designation, the second part relates to the machine code generation. This part begins with a traversal for the part's features. The aim is to search for the previously defined attributes in order to extract their data. By utilizing these data, the G-code is generated. First, it is formatted and added to a window for review. Next, it is stored to a standard text file document. The code generation and storage processes are triggered by the "Generate Machine Code" button. During the code generation, besides the machining parameters, the hole type, as well as the selected drilling strategy is being considered. Therefore, the appropriate drilling operations are being selected based on these factors. For instance, if a deep hole is identified and the multiple-drills strategy is selected, the peck drilling method will be considered. In contrast, if the one-drill strategy is selected, then the code will assume that the machine power is

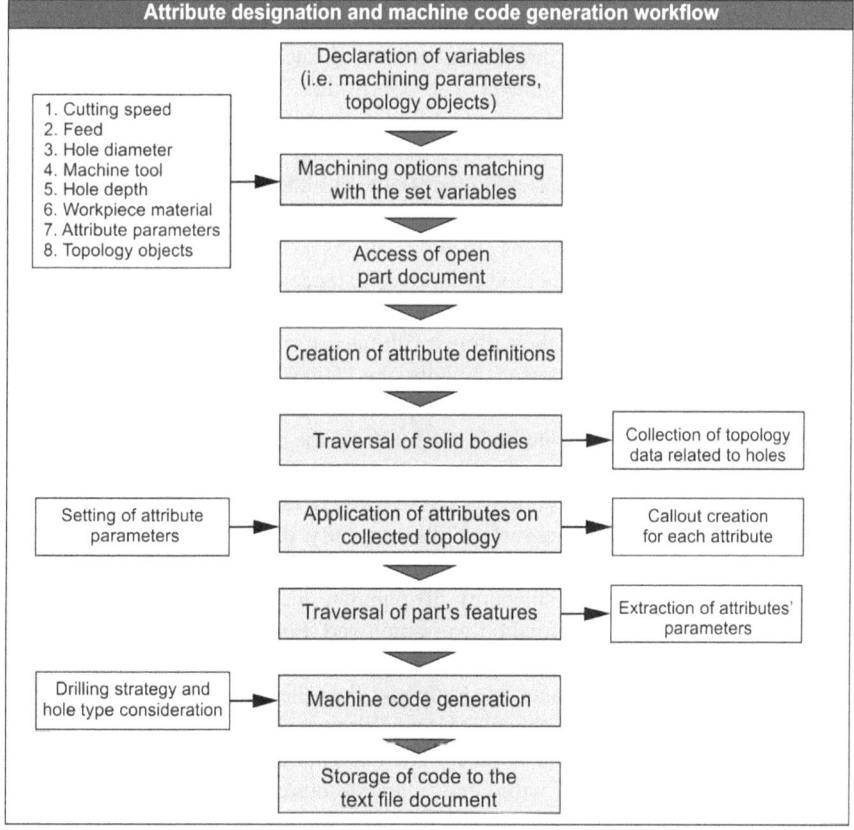

Fig. 5.2 Process flow for the design features identification and attributes designation

sufficient, and a long tool is available. To achieve this, multiple conditional structures were employed. Moreover, the necessary G and M commands, such as "G00" for the rapid tool movement and "M06" for the tool change, were utilized to properly form the code's template. It should be noted that the generated code does not use specific machine related code structure, such as FANUC, HAAS or HEIDENHAIN control. Therefore, minor adaptations are needed prior to simulating the tool path with specific CNC control software.

5.2.3   Process Code

The VBA™ language was used to write the code. The most important code snippet, related to the attribute designation, is presented in the Appendix section. The complete code is divided into two sub-procedures, each corresponding to a specific function. These procedures are triggered when the user clicks either the "Generate Machining Data" or "Generate Machine Code" button in the UI.

The next API methods, "DefineAttribute" and "AddParameter", are used to create an attribute definition for the machining parameters such as speed, feed, hole position coordinates, depth, and diameter. The "GetBodies2" method retrieves the solid bodies that are available in the open part document, while employed in a loop, which traverses the solid bodies. Methods "GetFirstFace" and "GetSurface" are utilized to locate the first face of the model and identify the cylindrical faces respectively. The "GetFirstFace" finds the first face, regardless of a cylindrical one or not. If this is the case however, the "CylinderParams" method collects the necessary geometrical data. Next, the "CreateInstance5" method creates an instance of the attribute according to the gathered data. Finally, the "CreateCallout2" and "CreateSelectData" methods are responsible for the display of the callouts with the information, when the "Generate Machining Data" button is pressed. Finally, the "GetNextFace" method is used to identify and select the next face in the body, for the attribute application. Table 5.1 lists the aforementioned API methods used to form the code of the first procedure.

The second procedure regards the generation of the CNC code. Once the attributes have been designated, a loop is performed to traverse the features. In a similar way to the "GetFirstFace" and "GetSurface" functionality, the "FirstFeature" and "GetSpecificFeature2" methods are used to get the first design feature of the model and the specific type of feature (attribute). To populate the preview window with the G-code lines, the "AddItem" method was employed. Finally, the "GetNextFeature" method is necessary for the next feature acquisition in the part document. In the case that the user selects a material group and thus the automatic selection of the cutting parameters is enabled, the "GetMaterialPropertyName2" method is required to identify the material of the workpiece. The realization of the automatic condition selection process is achieved with conditional structure. The same applies for the identification of the appropriate drilling methods. Table 5.2 presents the most basic API methods required for the machine code generation by identifying features that have designated attributes and thus are needed for the code production.

Table 5.1   API methods for the definition of the identified hole attributes [20]

API method	Explanation
DefineAttribute	Creates an attribute definition
AddParameter	Adds a parameter to the attribute definition using the default value and specified name
GetBodies2	Gets the bodies in the part
GetFirstFace	Finds the first face in a body and returns this face
GetSurface	Gets the surface referenced by this face
CylinderParams	Gets the parameters of a cylindrical surface
CreateInstance5	Creates an instance of this attribute on the specified object and its options
GetParameter	Gets the specified parameter on this attribute
SetDoubleValue2	Sets the double or integer value of a named configuration option parameter
CreateCallout2	Creates a callout for the selected object
CreateSelectData	Creates a ISelectData object to use as argument with Select methods
GetNextFace	Gets the next face in the body

Table 5.2   API methods for the machine code generation based on the design features [21]

API method	Explanation
FirstFeature	Gets the first feature in the open document
GetSpecificFeature2	Gets the interface for the selected feature
GetMaterialPropertyName2	Gets the names of the material database and the material for the specified configuration
AddItem	Adds the specified advanced component selection criterion to the created list
GetDoubleValue	Gets an attribute value of the double type
GetNextFeature	Gets the next feature in the part

Figure 5.3 illustrates the diagram for the selection of the basic drilling operations. The identification of the hole parameters is the first step, since it is important to obtain the hole diameter, depth, and type. Next, the collected data are matched to certain categories regarding the diameter, the depth, and the type. Therefore, hole size corresponds to micro holes (less than 1 mm in diameter), small to medium (between 1 and 20 mm), and large holes (more than 20 mm in diameter). The depth corresponds to either shallow holes or deep holes (depth-to-diameter ratio of at least 10:1). And type denotes the counterbore (counter) or countersink (chamfer) holes. Blind or through holes do not affect the decisions, however the depth can be simply compared to the part thickness for reference. The operation cycle is then selected according to the combination of the hole. For instance, a deep hole would require the peck drilling cycle, while at the same time, a chamfer would be required to be

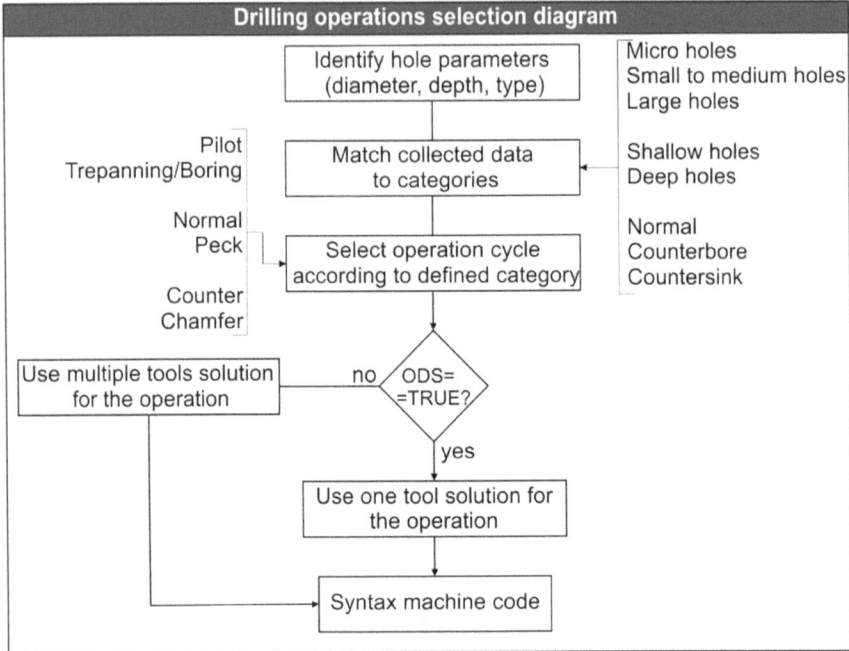

Fig. 5.3 Decision diagram for the drilling operations selection

formed also, if the hole is a countersink one. Moreover, a pilot drilling operation would be necessary as well.

Summarizing, in the case of a deep countersink hole, the pilot drilling and peck drilling operations would be required, in addition to the chamfer creation. Finally, before the code can be produced, a final selection is made according to whether the user have chosen the one-drill solution or not. In the case that the one-drill strategy is selected, the application would consider that an appropriate tool is available, and the hole can be made with solely one tool. Nowadays manufacturers build tools that are able to perform a pilot hole, cut the necessary hole, and create a chamfer at the same time. In the other case, the program assumes that all operations are carried out normally, each with the different tool.

5.3   Results

To test the developed tool, a sample part with two holes was used. Figure 5.4 illustrates the three steps that take place during the process. It is noted that part must be a parametrically designed one, in order for the features to be recognizable. For the testing, the native format of SolidWorks™ was used. However, standard file formats such as IGES (Initial Graphics Exchange Specification) and Parasolid can also be

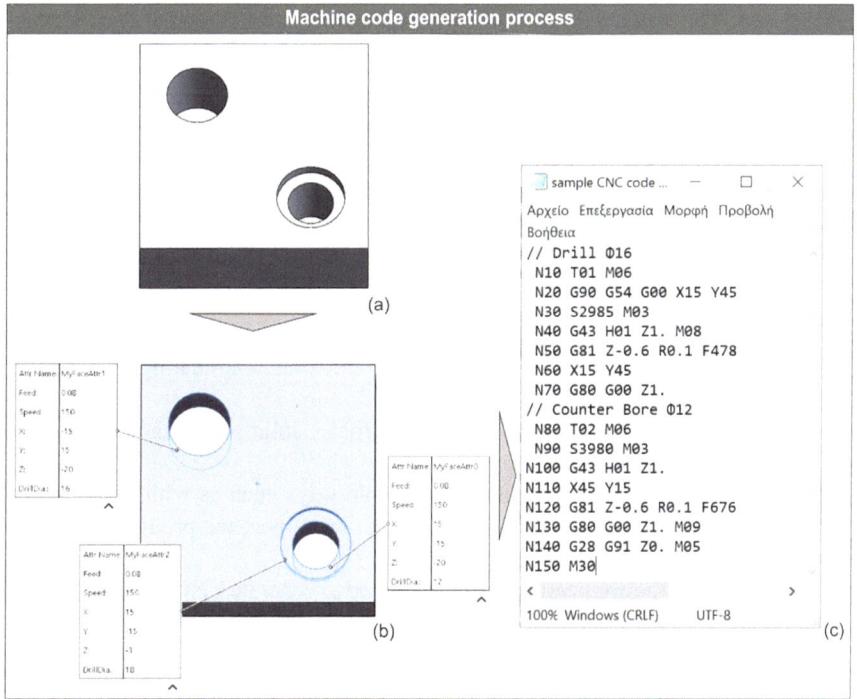

Fig. 5.4 Machine code generation for the drilling of sample part

used for this purpose, since the design features related to holes can be easily identified. Moreover, the user should take care of the way the part is designed, in terms of the coordinate system. Therefore, to ensure that the depth direction of the holes is aligned with the part's extrusion direction, the model should be designed with this in mind.

Figure 5.4a shows a sample workpiece, parametrically designed in SolidWorks™, with two holes: one normal through hole and one counterbore. Figure 5.4b shows the created callouts for the two holes, which contain information that is required for the generation of the CNC code such as the position of the hole, the depth, the diameter, and the machining parameters (speed and feed). The callouts are generated after the hole identification process takes place, which involves the scanning of the cylindrical surfaces found the model and the identification of those related to hole or extruded-cut features. Finally, Figure 5.4c displays the text file that contains the generated code, as shown in the preview window. Users have the opportunity to check the validity of the code by simulating it with the aid of a CAM system or a toolpath simulator. The specific code was tested with a free online code viewer software, after some minor modifications were made to match the software's format.

5.4   Conclusions

In this work, the development of an automated CNC code generator tool is being demonstrated. The automation is achieved by employing the API of the SolidWorks™ CAD system, without the need for specialized software packages. The process is based on the concept of traversing the bodies of the part document, identifying and retrieving every circular edge on the solid bodies.

By testing and analyzing the developed tool, the following conclusions can be drawn:

- The developed tool provides an alternative method for generating CNC codes for drilling, by using a standard CAD system. Thus, eliminating the need for expensive dedicated CAM systems.
- It enables the direct generation of G-code from a solid part without converting the model to a two-dimensional document.
- The generated code can be edited in multiple ways such as with a simple text editor or by importing it to a CAM software. Moreover, the produced code can be imported into simulation software.
- The usability of the application can be extended to generate CNC codes for similar processes as well.

Finalizing, the programming of CAD systems can be used to automate other manufacturing-related tasks as well, leading to reduced production time and cost of a product. Consequently, productivity can improve, and time management can become more effective.

Appendix

```vba
Private Sub UserForm_Activate()
    Set swApp = Application.SldWorks
    Set swModel = swApp.ActiveDoc
'Acquire part thickness
    Set swDim = swModel.Parameter("D1@Boss-Extrude1")

    Debug.Print "File = " & swModel.GetPathName
    Debug.Print "  Full name = " & swDim.FullName
    Debug.Print "  Name = " & swDim.Name

    vConfigNames = swModel.GetConfigurationNames
    vValue = swDim.GetSystemValue3(swThisConfiguration, _
(vConfigNames))

    Debug.Print "  System value = " & vValue(0) * 1000#; _
"" & " mm"
'-----------------------------------------------------
    If swModel Is Nothing Then
        MsgBox "Please open a part document"
        Exit Sub
    End If

    If swModel.GetType <> swDocPART Then
        MsgBox "Please open a part document"
        Exit Sub
    End If
        'Create Attribute Definition
    Set swAttDef = swApp.DefineAttribute _
                        ("pubMyFaceAttributeDef")
  swAttDef.AddParameter "FeedRate", _
                    SwConst.swParamTypeDouble, 10, 0
    swAttDef.AddParameter "SpeedRate", _
                    SwConst.swParamTypeDouble, 20, 0
    swAttDef.AddParameter "XPos", _
                    SwConst.swParamTypeDouble, 0, 0
    swAttDef.AddParameter "YPos", _
                    SwConst.swParamTypeDouble, 0, 0
    swAttDef.AddParameter "ZPos", _
                    SwConst.swParamTypeDouble, 0, 0
    swAttDef.AddParameter "Depth", _
                    SwConst.swParamTypeDouble, 0, 0
    swAttDef.AddParameter "HoleDiameter", _
                    SwConst.swParamTypeDouble, 0, 0
    swAttDef.Register
    End Sub
```

```vb
Private Sub cmdGenerateMachiningData_Click()

'Traverse the body to find all cylindrical surfaces
    Set CalloutHandler = New AttCalloutHandler
    Set CalloutCollection = New Collection
    Set swPart = swModel
    If Not swPart Is Nothing Then
        retval = swPart.GetBodies2(swSolidBody, True)
        For i = 0 To UBound(retval)
            Dim j As Integer
            j = 0
            Set swFace = retval(i).GetFirstFace
            Do While Not swFace Is Nothing
                Dim swSurface As surface
                Set swSurface = swFace.GetSurface
        'If the current face surface type is cylindrical
        'then gather it's data and designate an attribute
                If swSurface.IsCylinder Then
                    Dim cylParams As Variant
                    cylParams = swSurface.CylinderParams
        'Create an instance of the attribute
Set swAtt = swAttDef.CreateInstance5 _
(swModel, swFace, "MyFaceAttribute-" & j, 0, _
SwConst.swAllConfiguration)
            'Set each parameter value of the attribute using
the data from the cylindrical surface
                If Not swAtt Is Nothing Then
                    Set swAttParam = swAtt.GetParameter _
                        ("Feed(mm/min)")
```

```vb
            If Not swAtt Is Nothing Then
          Set swAttParam = swAtt.GetParameter _
                     ("Feed(mm/min)")
bRet = swAttParam.SetDoubleValue2 _
                 (feed.Value, _
               SwConst.swAllConfiguration, "")
       Set swAttParam = swAtt.GetParameter _
                  ("Speed(m/min)")
bRet = swAttParam.SetDoubleValue2(speed.Value, _
               SwConst.swAllConfiguration, "")
       Set swAttParam = swAtt.GetParameter _
                  ("X")
bRet = swAttParam.SetDoubleValue2 _
                 (cylParams(0) / 0.001, _
               SwConst.swAllConfiguration, "")
       Set swAttParam = swAtt.GetParameter _
                  ("Y")
bRet = swAttParam.SetDoubleValue2 _
                 (cylParams(1) / 0.001, _
               SwConst.swAllConfiguration, "")
       Set swAttParam = swAtt.GetParameter _
                  ("Z")

        bRet = swAttParam.SetDoubleValue2 _
                       (cylParams(2) / 0.001, _
                     SwConst.swAllConfiguration, "")
           Set swAttParam = swAtt.GetParameter _
                       ("HoleΦ")
        bRet = swAttParam.SetDoubleValue2 _
                      ((cylParams(6) / 0.001) * 2, _
                     SwConst.swAllConfiguration, "")
           Set swAttParam = swAtt.GetParameter _
                       ("Depth")
        bRet = swAttParam.SetDoubleValue2 _
                      (cylParams(1) / 0.001, _
                     SwConst.swAllConfiguration, "")

'Display which faces got attributes, use the callout

          Dim swSelMgr As SldWorks.SelectionMgr
          Set swSelMgr = swModel.SelectionManager
          'Create a callout object
          Dim swCallout As SldWorks.Callout
          Set swCallout = swSelMgr.CreateCallout2 _
                    (7, CalloutHandler)
```

```vb
                 swCallout.TargetStyle = _
        SwConst.swCalloutTargetStyle_e. _
                           swCalloutTargetStyle_Circle
                           swCallout.Label2(0) = "Attr Name"
                           swCallout.Value(0) = "MyFaceAttr" & j
                           swCallout.Label2(1) = "Feed"
                           swCallout.Value(1) = feed.Value
                           swCallout.Label2(2) = "Speed"
                           swCallout.Value(2) = speed.Value
                           swCallout.Label2(3) = "X"
                           swCallout.Value(3) = _
                                   Round(cylParams(0) / 0.001, 2)
                           swCallout.Label2(4) = "Y"
                           swCallout.Value(4) = _
                                   Round(cylParams(1) / 0.001, 2)
                           swCallout.Label2(5) = "Z"
                           swCallout.Value(5) = _
                                   Round(cylParams(2) / 0.001, 2)
                           swCallout.Label2(6) = "DrillΦ"
                           swCallout.Value(6) = _
                                   (cylParams(6) / 0.001) * 2
                           CalloutCollection.Add swCallout

            Dim SelData As SldWorks.SelectData
            Set SelData = swSelMgr.CreateSelectData
                    SelData.Callout = swCallout

                        SelData.Mark = 0
                        swFace.Select4 True, SelData
                        swCallout.ValueInactive(0) = True
                        Set swAtt = Nothing
                    End If
                    j = j + 1
                End If
                Set swFace = swFace.GetNextFace
              Loop
            Next i
        End If
        Set swPart = Nothing
End Sub
```

References

1. Favi C, Mandolini M, Campi F, Germani M (2021) A CAD-based design for manufacturing method for casted components. Procedia CIRP 100:235–240. https://doi.org/10.1016/j.procir.2021.05.061
2. Matta AK, Raju DR, Suman KNS (2015) The integration of CAD/CAM and rapid prototyping in product development: a review. Mater Today Proc 2:3438–3445. https://doi.org/10.1016/j.matpr.2015.07.319
3. Balic J (2006) Intelligent CAD/CAM systems for CNC programming–an overview. Adv Prod Eng Manag 1:13–22

4. König O, Wintermantel M (2004) CAD-based evolutionary design optimization with CATIA V5. Weimarer Optimierungs und Stochastiktage 10:1–30
5. Alonso J, Martins JRRA, Reuther J, et al (2003) High-fidelity aero-structural design using a parametric CAD-based model. In: 16th AIAA computational fluid dynamics conference
6. Bordegoni M (2011) Product virtualization: an effective method for the evaluation of concept design of new products. In: Bordegoni M, Rizzi C (eds) Innovation in product design: from CAD to virtual prototyping. Springer, London, London, pp 117–141
7. Rama Murthy S, Mani M (2012) Design for sustainability: the role of CAD. Renew Sustain Energy Rev 16:4247–4256. https://doi.org/10.1016/j.rser.2012.03.009
8. Cera CD, Braude I, Kim T et al (2005) Hierarchical role-based viewing for multilevel information security in collaborative CAD. J Comput Inf Sci Eng 6:2–10. https://doi.org/10.1115/1.2161226
9. Tzotzis A, Tsagaris A, Tapoglou N, Kyratsis P (2023) High-precision CAD-based simulation for turning considering tool microgeometry. Int J Mechatron Manuf Syst 16:83–95. https://doi.org/10.1504/IJMMS.2023.132023
10. Chacón JM, Bellido JC, Donoso A (2014) Integration of topology optimized designs into CAD/CAM via an IGES translator. Struct Multidiscip Optim. https://doi.org/10.1007/s00158-014-1099-6
11. Tapoglou N, Efstathiou C, Tzotzis A, Kyratsis P (2023) Study of the topography of face milled surfaces using CAD-based simulation. In: Kyratsis P, Manavis A, Davim JP (eds) Computational design and digital manufacturing. Springer International Publishing, Cham, pp 159–166
12. Tzotzis A, García-Hernández C, Huertas-Talón JL, Kyratsis P (2020) CAD-based automated design of FEA-ready cutting tools. J Manuf Mater Process 4:1–14. https://doi.org/10.3390/jmmp4040104
13. Gella-Marín R, Tzotzis A, García-Hernández C, et al (2021) CAD software integration with programming tools for modelling, measurement and verification of surfaces. In: Experiments and simulations in advanced manufacturing, pp 91–116
14. Kyratsis P (2020) Computational design and digital manufacturing. Int J Mod Manuf Technol 12:82–91
15. Ribeiro-Carvalho S, Horovistiz A, Davim JP (2021) Material model assessment in Ti6Al4V machining simulations with FEM. Proc Inst Mech Eng Part C J Mech Eng Sci 235:5500–5510. https://doi.org/10.1177/0954406221994883
16. Kyratsis P, Tzotzis A, Manavis A (2021) Computational design and digital fabrication. In: Kumar S, Rajurkar KP (eds) Advances in manufacturing systems. Springer Singapore, Singapore, pp 1–16
17. Manavis A, Tzotzis A, Tsagaris A, Kyratsis P (2022) A novel computational-based visual brand identity (CbVBI). Machines
18. Tsagaris A, Polychroniadis C, Tzotzis A, Kyratsis P (2024) Cost-effective robotic arm simulation and system verification. Int J Intell Syst Appl 16:1–12. https://doi.org/10.5815/ijisa.2024.02.01
19. Tzotzis A, Garcia-Hernandez C, Huertas-Talon J-L et al (2017) Engineering applications using CAD based application programming interface. In: MATEC web of conferences, pp 1–7
20. Tzotzis A, Efkolidis N, García-Hernández C, Kyratsis P (2024) CAD-based automated USV hull design. In: Guxho G, Kosova Spahiu T, Prifti V et al (eds) Proceedings of the joint international conference: 10th textile conference and 4th conference on engineering and entrepreneurship. Springer Nature Switzerland, Cham, pp 354–362
21. Tzotzis A, Manavis A, Efkolidis N, Kyratsis P (2021) CAD-based automated G-code generation for drilling operations. Int J Mod Manuf Technol 13:177–184. https://doi.org/10.54684/ijmmt.2021.13.3.177

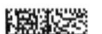